Synthesis Lectures on Wave Phenomena in the Physical Sciences

Series Editor

Sanichiro Yoshida, Department of Chemistry and Physics, Southeastern Louisiana University, Hammond, USA

The aim of this series is to discuss the science of various waves. An emphasis is laid on grasping the big picture of each subject without dealing formalism, and yet understanding the practical aspects of the subject. To this end, mathematical formulations are simplified as much as possible and applications to cutting edge research are included.

Rhett Allain

Modeling Waves
with Numerical Calculations
Using Python

 Springer

Rhett Allain
Southeastern Louisiana University
Hammond, LA, USA

ISSN 2690-2346   ISSN 2690-2354 (electronic)
Synthesis Lectures on Wave Phenomena in the Physical Sciences
ISBN 978-3-031-78290-9   ISBN 978-3-031-78291-6 (eBook)
https://doi.org/10.1007/978-3-031-78291-6

This Springer imprint is published by the registered company Springer Nature Switzerland AG
The registered company address is: Gewerbestrasse 11, 6330 Cham, Switzerland

If disposing of this product, please recycle the paper.

Contents

Introduction to Numerical Calculations: A Mass on a Spring

1.1 Introduction

It's one of the reoccurring mathematical models in physics—waves. Of course you can have a wave on a string or even a two dimensional surface like the head of a drum. Light is also an electromagnetic wave also. In quantum mechanics, we can model the motion of particle with the Schrodinger wave equation (which isn't the same thing as a wave equation).

In one dimension, we can write the wave equation as the following:

$$\frac{d^2y}{dt^2} = v^2\frac{d^2y}{dx^2}$$

Although there are analytical solutions to this equation, there are also countless situations for which the only solution is numerical.

Of course that leaves two important questions. What exactly is a numerical solution? How do you create a numerical solution? That's exactly what will be covered in this book. I'm going to step you through everything you need to know to create numerical solutions of the wave equation.

What do you need to know to start this process? I will assume that you have a basic understanding of both physics and differential equations (like the wave equation). However, I will NOT assume you have any previous experience with programming. Yes, in order to create a numerical solution we will need to write some computer code. In this case, we are going to use an online version of python called Web VPython. You can access this using a web browser at https://www.glowscript.org. I'm going to walk you through all the steps starting from scratch.

© The Author(s), under exclusive license to Springer Nature Switzerland AG 2025
R. Allain, *Modeling Waves with Numerical Calculations Using Python*, Synthesis Lectures on Wave Phenomena in the Physical Sciences,
https://doi.org/10.1007/978-3-031-78291-6_1

1.2   A Physics Problem

In order to really understand numerical calculations, we need to start with a simple example that can be solved both analytically and numerically. This is a very common problem, so there's a chance you've already solved it but let's do it again.

Suppose a mass (m) is on a horizontal frictionless table and connected to a spring with a spring constant (k). The mass is pulled to the side such that it has a position x_0 and released from rest.

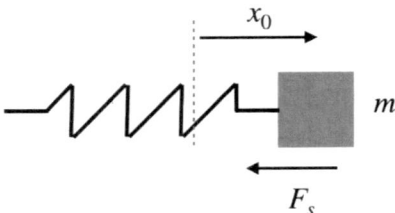

Where the spring force (in the x-direction) is:

$$F_s = -kx$$

Since this is the only force, Newton's second law becomes:

$$F_s = -kx = ma_x = m\frac{d^2x}{dt^2}$$

Solving for $\frac{d^2x}{dt^2}$:

$$\frac{d^2x}{dt^2} = -\frac{k}{m}x$$

One way to solve this differential equation is to just guess a solution. We want a function for x that has the second derivative equal to a negative constant multiplied by that function. Here's one (of many) possible solutions.

$$x(t) = A\cos\omega t$$

Taking the derivative (with respect to t) twice gives:

$$\frac{d^2x}{dt^2} = -A\omega^2\cos\omega t$$

So that this is a solution if $\omega = \sqrt{\frac{k}{m}}$ and $A = x(0)$. But this is the solution you may have already seen in any number of physics courses.

1.3   Numerical Solution

Let's solve this problem another way. Suppose I take the mass connected to a spring and pull it slightly so that it is no longer at its equilibrium position. When the mass is released, there is a force pulling on it that will change its motion. If this spring force was constant we could use it to calculate the constant acceleration. With a constant acceleration it would be simple to find the change in velocity and the change in position of the mass. However, the spring force is NOT constant but rather changes as the mass moves closer to the equilibrium position.

One way to deal with this changing force problem is to break the motion of the mass into very short time intervals. During these intervals we can approximate the force as having a constant value. With a shorter time interval, this approximation is closer to the actual force. Suppose that we have a time interval of $\Delta t = 0.01s$ then we can write the acceleration as:

$$a = \frac{F}{m} = \frac{\Delta v}{\Delta t}$$

Now we can label the velocity at the beginning of this time interval as v_1 and the velocity at the end as v_2. Putting these in for Δv and solving for v_2 gives:

$$\frac{F}{m} = \frac{v_2 - v_1}{\Delta t}$$

$$v_2 = v_1 + \left(\frac{F}{m}\right)\Delta t$$

Next we can make another crazy assumption—that the velocity is constant during this time interval (which is clearly not true since we just calculated the change in the velocity). However, if the velocity was constant then we could set it equal to the rate of change of position where x_1 is the position at the beginning of the time interval and x_2 is at the end.

$$v = \frac{\Delta x}{\Delta t} = \frac{x_2 - x_1}{\Delta t}$$

$$x_2 = x_1 + v\Delta t$$

With these approximations, we have found the new position and velocity of the mass after the 0.01 s time interval. We can repeat this calculation for the next time interval (from 0.01 to 0.02 s). If we want to look at the motion over a full 1 s, these calculations would need to be repeated 100 times. No one wants to do that. Instead, we can use a computer for this task. That's exactly what we will do.

1.4   Web VPython

Like I said before, I'm not going to expect that you have any previous knowledge about programming in python (or any other language). Python has a fairly simple syntax, but it can also be quite powerful. In this case we can use an online version of python called Web VPython (https://www.glowscript.org). This means that a user won't have to install anything to get this to function. Web VPython also has some other very useful features:

- Built in mathematical functions as well as vector operations.
- Simple methods for graphing (including animated graphs).
- Tools to build 3D visualizations for the motions of objects (we will get into this later).

Let's start off with some very basic operations in python. Here's a very simple program so that I can point out some important features.

```
 1  Web VPython 3.2
 2
 3  x = 2
 4  y = 3
 5  z = x + y
 6  print("z = ",z," meters")
 7  X = 2.2 #meters
 8  print("X = ",X," m")
 9  y = y + 1
10  print("y = ",y," m")
11
12
```

Comments:

- Line 1: This line is needed so that the code knows which version of the software to run. It's best to just leave this line alone.
- Line 3: Here we are declaring the variable "x" to be equal to the value of 2. You don't need to declare the type of variable—python figures that because it's set to a number.
- Line 5: If you have two variables, you can do mathematical operations and make that equal to a new variable.

- Line 6: The print statement can be a combination of strings (like "z = ") and variables. Be sure to separate these with a comma.
- Line 7: Python is case sensitive. X is different than x. I suggest you don't use both of those variables in the same program since humans are not as sensitive to case.
- Line 7: Also in this line you can see a comment. Using the "#" indicates that everything following is ignored by python (but not by the human reader). This can be a useful way to incorporate units into the calculation since Python just wants to deal with numerical values.
- Line 9: This is a great example to show that "=" does not mean "algebraic equal". Instead, the equal sign means "make equal to". In this case, it takes the value of y (which is 3) and adds 1 to it and then sets the value of y equal to this number. So, this line actually changes the value of y. We will use this quite a bit.

When you run the code, you get output that look like this.

```
z =  5 meters
X =  2.2 m
y =  4 m
```

We need one more very useful concept in python—loops. Here is some code:

```
1 Web VPython 3.2
2
3 t = 0
4 dt = 0.1
5
6 while t<1:
7       print("t = ",t)
8       t = t + dt
9
```

In python, a while loop repeats everything after a colon that is tab indented. In this case that means as long as the value of t is less than 1. Since t starts off with a value of zero, it will least go through the code once. Notice that in line 8 the time is increased by a value of 0.1 so that eventually the value of t will be greater than 1 and the python code will move on to whatever is after the loop (in this case, that's nothing). The output looks like this:

```
t = 0.6
t = 0.7
t = 0.8
t = 0.9
t = 1
```

That's what we expected.

Now we are ready to model this mass on a spring. Just to be clear, we need to do the following during each time interval.

- Calculate the force from the spring (this depends on the position of the mass).
- Update the velocity of the mass.
- Update the position of the mass.
- Update time.

Here's what this looks like in python.

```
1  Web VPython 3.2
2
3  m = 0.1
4  k = 10
5  t = 0
6  dt = 0.01
7  x = 0.01
8  v = 0
9
10 while t<1:
11     F = -k*x
12     v = v + (F/m)*dt
13     x = x + v*dt
14     t = t + dt
15
16 print("x = ",x," m")
17
```

Some important comments:

- We need numerical values. This is a numerical calculations so you need numbers. I just picked some arbitrary values for m and k.

- Since the code updates the velocity and the position, we need to have initial values for these variables. Notice that we don't have to call them v_0 and x_0 but rather just v and x. These are just the starting values—we are going to change them in the code.
- The code prints the final value of the position—but that's not very insightful. Maybe we need to create a graph.

1.5   Graphs in Web VPython

Let's add a graph to this simple harmonic oscillator code. Really, it only involves three new lines. Here's the modified code (and I will then give comments on the additional lines).

```
 3 g1 = graph(title="Mass on Spring", xtitle="t [s]",
 4 ytitle="x [m]", width=400, height = 200)
 5 f1 = gcurve(color=color.blue)
 6 m = 0.1
 7 k = 10
 8 t = 0
 9 dt = 0.01
10 x = 0.01
11 v = 0
12
13 while t<1:
14     F = -k*x
15     v = v + (F/m)*dt
16     x = x + v*dt
17     f1.plot(t,x)
18     t = t + dt
```

Here are the three important lines:

- Line 3: This makes the graph axes using the built in graph() function. In this case, I gave it the name "g1"—but you can give it any reasonable name. The graph function has properties such as the x- and y-titles as well as the size. Note: I wrapped the code around to line 4 (which is allowed).
- Line 5: Here we define the actual curve to be plotted. I give this the name "f1" and it's a type of curve called a gcurve (which plots points with a line connecting them). It also has several properties, but the one I always use is the color.
- Line 17: In the loop, I need to plot a point on the graph. This is accomplished by calling f1.plot and then passing the horizontal and vertical variables (in this case t and x).

That's it. Here's the output.

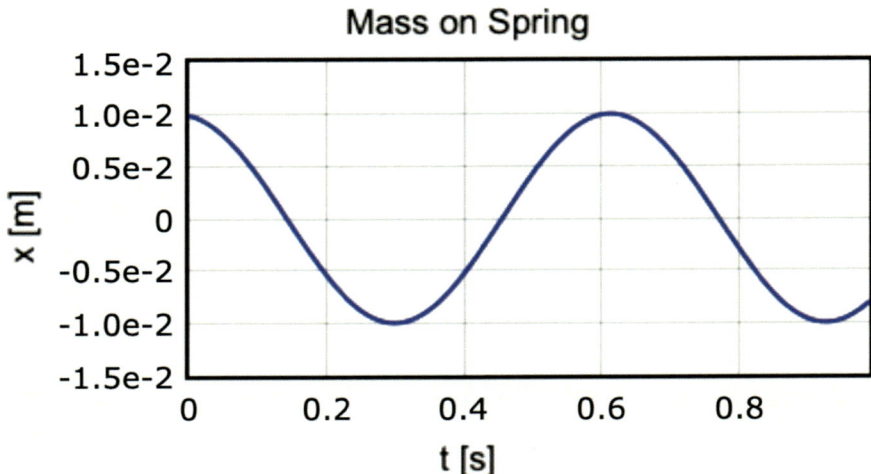

That's exactly what we expected. A plot that looks like a cosine function. However, just a reminder, it is NOT a cosine function—it's just data. The numerical calculation doesn't give us an analytical solution to the problem, it just gives us numbers. In this case, those numbers seem to be the same thing that the cosine function would give us and that's a good thing. It suggests that the numerical calculation works.

1.6   3D Visualization

There is one more part of Web VPython that we can use for this basic simple harmonic oscillator. The "V" part of Web VPython stands for visual. Yes, we can make a 3D visualization of a mass on a spring. While it's true that the graphical output for a mass on a spring gives us what we really need, a visualization can be also be very useful in some situations.

Web VPython has some built-in 3D objects. But before we get to 3D objects, we need to think about how to deal with vectors in Web VPython. There is a built in vector class as well as some vector operations. Consider the following code:

```
 1  Web VPython 3.2
 2
 3  A = vector(1,2,3)
 4  B = vector(0.1,-.5,1.1)
 5  C = A + B
 6  print("C = ",C)
 7  a = 0.1
 8  print("aA = ",a*A)
 9  print("|A| = ",mag(A))
10  print("A-hat = ",norm(A))
11  print("A*B = ",dot(A,B))
12  print("AxB = ",cross(A,B))
13
```

Here we create two variable that are vectors (A and B). Note that using the syntax A = vec(1,2,3) works as well if writing out "vector" is too much. In line 5, you can see that vector addition works. The other operations are scalar multiplication (line 8), magnitude of a vector (line 9), unit vectors (line 10), dot product (line 11), and finally cross product (line 12).

Now we can make a very simple program with a 3D object. It looks like this.

```
 1  Web VPython 3.2
 2  canvas( background=color.white)
 3  ball = sphere(pos=vector(1,2,0), radius=0.3, color=color.red)
 4
 5
```

Note: In most cases, I'm going to add code like you see in line 2. It's not important, but it changes the background color from the default black to white. This is only to make the output fit better into this text format.

The rest of the code only has one line. Here we create a sphere object and give it the name "ball". The sphere() is a built in function that has several properties—but in this case we will just assign the position (pos), the radius, and the color. The position is the vector location (in 3D space) for the center of the ball. Here's what the output looks like.

It's actually a 3D representation. Using the mouse, you can interact with this environment by zooming and panning the "camera". Although there are many available 3D objects, we can create an oscillating mass on a spring with just a couple more.

There is the box(). This, not surprisingly, creates a 3D box. The important parameters are the position (pos) that give the vector location of the center of the box. The size parameter is a vector value that sets the x, y, and z lengths of the box. All objects also have the color attribute. If you do not specify a color value, it will be white. You can still see a white object on a white background because of its shadow.

The last object we need is a helix to visualize the spring. The one thing that's quite different about the helix (and the cylinder object that will be used later) is that the position is the location of one end of the helix. The orientation of the object is defined by the axis property. This is vector from the position of the object to the other end of the object.

Finally, it's important to note that you can access the attributes of an object. The location of the ball would be ball.pos. If you change the value of ball.pos, the ball will move in the 3D environment. This is part of the animation process.

OK, let's make a mass oscillating on a spring with these three objects. Here is the code.

```
1  Web VPython 3.2
2  canvas(background=color.white)
3  ball = sphere(pos=vector(0.01,0,0), radius=0.002, color=color.red)
4  wall = box(pos=vector(-0.02,0,0), size=vector(0.001,0.006,0.006))
5  spring = helix(pos=wall.pos, axis=ball.pos-wall.pos,
6  thickness=0.0002,radius=0.001)
7  m = 0.1
8  k = 10
9  t = 0
10 dt = 0.01
11 x = 0.01
12 v = 0
13
14 while t<3:
15     rate(100)
16     F = -k*x
17     v = v + (F/m)*dt
18     x = x + v*dt
19     ball.pos = vector(x,0,0)
20     spring.axis=ball.pos-wall.pos
21     t = t + dt
22
```

This gives an animated output, here's a image of that (but it's not animated on paper).

Here are the important details regarding the code.

- Line 1: This creates the ball. Nothing new here.
- Line 4: I picked some values for the location and size of the wall so that it looked nice.
- Line 5: Here is the spring. Notice that for the position I set it equal to wall.pos and the axis is from the wall to the ball so that would be ball.pos-wall.pos.
- Line 15: Since we want to animate the ball on the spring, we need to add a a rate statement "rate (100)". This tells the loop to not do any more than 100 iterations every second. Since we have a time step of 0.01 s, a loop while t < 3 would take 3 s. This means the animation will run in "real time".
- Line 19: Here I am updating the position of the ball. Note that this is not the best way to do it, but I wanted to use the previous code that's a bit simpler.
- Line 20: If the ball moves, the spring needs to also change.

1.7 A Better Model for the Spring Force

Although the simple code above gives a nice output for a mass oscillating on a spring, it only works in 1 dimension and it has a spring with an unstretched length of 0 m. That's not very realistic.

Instead, let's imagine that I have a spring with an unstretched length L_0 and spring constant k. When it is stretched or compressed, I can define it's size with the vector $\vec{L}$ that goes from one end to the other of the spring. Here's a diagram.

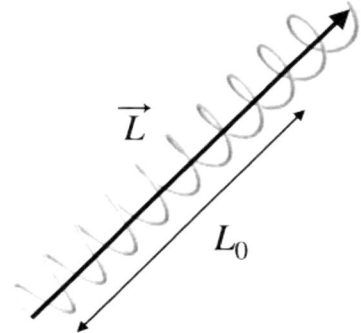

In this case, the stretch of the spring would be $\left|\vec{L}\right| - L_0$. Notice that the length is a vector but the unstretched length is a scalar. In order to find the magnitude of the stretch, we need to first find the magnitude of the length vector. If we multiply this stretch by the spring constant, it will give us the magnitude of the force the spring exerts. However, we want more than the magnitude of the vector—we want the vector force. We can fix this by multiplying by $\hat{L}$—a unit vector in the direction of $\vec{L}$. This gives the following expression for the force from a real spring.

$$\vec{F_s} = -k\left(\left|\vec{L}\right| - L_0\right)\hat{L}$$

Not only does this model take into account the unstretched length of the spring, it gives a vector value and allows the spring to move in any direction.

Let's use this model to rebuild the mass on a spring. Just to make it more interesting, we can model a mass hanging vertically with a downward pulling gravitational force. Since we are dealing with vectors values for the force, we should make a small modification to our numerical recipe. First, we can calculate the net force on the mass:

$$\vec{F_{net}} = m\vec{g} - k\left(\left|\vec{L}\right| - L_0\right)\hat{L}$$

where $\vec{g}$ is the gravitational field with a value of $\vec{g} = \langle 0, -9.8, 0\rangle$ N/kg. Second, instead of updating the velocity, it can be more general to think of the momentum principle.

$$\vec{F_{net}} = \frac{\Delta \vec{p}}{\Delta t}$$

where $\vec{P} = m\vec{v}$. This means that during the short time interval, we can update the momentum as:

$$\vec{p_2} = \vec{p_1} + \vec{F_{net}}\Delta t$$

Then, the position update step would be:

$$\vec{r_2} = \vec{r_1} + \left(\frac{\vec{p}}{m}\right)\Delta t$$

where $\vec{r}$ is the position of the mass. Here is the code for this motion.

```
1  Web VPython 3.2
2  canvas(background=color.white)
3
4  m = 0.03
5  L0 = 0.05
6  k = 10
7  g = vector(0,-9.8,0)
8  top = box(pos=vector(0,L0/2,0), size=vector(L0/2,L0/10,L0/2))
9  ball = sphere(pos=top.pos+vector(0,-L0,0),radius=0.005,
10 color=color.red, make_trail=True)
11 spring=helix(pos=top.pos, axis=ball.pos-top.pos, radius=0.002,
12 coils=10, thickness=0.001)
13 ball.p = m*vector(0,0,0)
14 t = 0
15 dt = 0.01
16
17 while t<4:
18     rate(100)
19     L = ball.pos-top.pos
20     Fnet = m*g-k*(mag(L)-L0)*norm(L)
21     ball.p = ball.p + Fnet*dt
22     ball.pos = ball.pos + ball.p*dt/m
23     t = t + dt
24     spring.axis = ball.pos-top.pos
25
```

Comments:

- Note that lines 9 and 11 are wrapped (so that they will fit).
- In line 10, the property "make_trail=True" means that the ball will leave a trail during animation.
- Line 13: I'm assigning the momentum of the ball as a property of the ball with ball.p— this is useful way to keep track of values.
- Line 19: notice that in the loop, I can calculate the vector value of L using the position of the top (the hanging point) and the box.
- Line 20: contains the gravitational force as well as the spring force.
- At the end of the loop (line 24) I need to remember to update the axis of the spring.

If you want, you could also add a graph to plot the vertical position as a function of time just like before. But here's something very new. Suppose that I start the ball with an initial position that is NOT right below the hanging point. In this case, the spring force will be pulling both up AND to the side. The motion of the ball will not be a plain simple harmonic oscillator, but something else. Here's a picture of the ball's motion in that case (using the trail so you can see what it did).

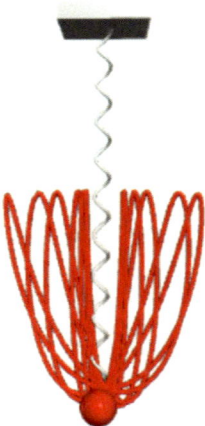

This shows the power of the 3D visualizations—especially as an animation.

Waves on a String and the Wave Equation

2

Before building a numerical model of a wave on a string, let's take a moment to create a mathematical model for wave motion. Suppose there is a horizontal string with a small vertical displacement. This displacement can then travel down the length of the string as a wave.

For our calculation, we can make the following assumptions and approximations.

- Parts of string only move up and down in the vertical direction.
- Deviation in the vertical direction is small compared to the length of the string.
- Even though parts of the string move up and down, the tension in the string remains constant.
- The gravitational interaction with the string is negligible.
- There are no frictional forces.

With that, let's start off by modeling the string as many individual masses connected by a massless string. It might look like this:

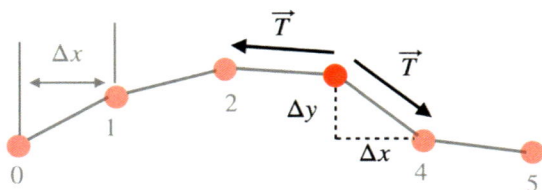

© The Author(s), under exclusive license to Springer Nature Switzerland AG 2025   15
R. Allain, *Modeling Waves with Numerical Calculations Using Python*, Synthesis
Lectures on Wave Phenomena in the Physical Sciences,
https://doi.org/10.1007/978-3-031-78291-6_2

Notice that each mass is the same horizontal distance from the other masses, a value of Δx. Yes, even though the masses can move up and down this distance will still be constant since the vertical motion is small (Δy).

Let's consider the motion of the element labeled "3" above. There are two forces acting on this mass. There's the tension pulling to the left due to the position of mass 2 and then there's the tension pulling to the right. For the tension pulling to the right, it must be in the same direction as the string (strings only pull in the direction of the string). Going from mass 3 to 4 has a change in vertical position of Δy. Using this change in vertical and horizontal position gives the following relationship.

$$\frac{T_y}{T} = \frac{\Delta y}{\Delta x}$$

We can only say this since Δy is small compared to Δx and the tension in the string is constant. Let's write Δy in terms of the y-values for mass 3 and 4 and solve for the vertical component of the tension.

$$T_y = T\frac{(y_4 - y_3)}{\Delta x}$$

This calculation can then be repeated for the tension pulling to the left. However, the direction of this tension is in the opposite direction of Δx since it's the left string. Adding these two forces we can get the net vertical force on the 3rd mass.

$$T_{net\text{-}y} = T\frac{y_4 - y_3}{\Delta x} - T\frac{y_3 - y_2}{\Delta x} = T\left(\frac{y_4 + y_2 - 2y_3}{\Delta x}\right)$$

This can be generalized to the i-th element as:

$$T_{y\text{-net}} = T\left(\frac{y_{i+1} + y_{i-1} - 2y_i}{\Delta x}\right)$$

Now that we have the net force in the y-direction, we can use this with Newton's second law.

$$F_{net\text{-}y} = m_i a_{yi}$$

For the mass of this element, we can assume a uniform linear density (mass per length) of μ where:

$$\mu = \frac{m}{L}$$

With m the total mass of the string and L the length. This means that the mass of one element would be $\mu \Delta x$. Finally, we can write the acceleration as the second derivative of the y-position with respect to time.

$$a_{yi} = \frac{d^2 y_i}{dt^2}$$

Putting this all together gives the following:

$$F_{net-y} = \mu \Delta x \frac{d^2 y_i}{dt^2}$$

$$T\left(\frac{y_{i+1} + y_{i-1} - 2y_i}{\Delta x}\right) = \mu \Delta x \frac{d^2 y_i}{dt^2}$$

$$\frac{T}{\mu}\left(\frac{y_{i+1} + y_{i-1} - 2y_i}{\Delta x^2}\right) = \frac{d^2 y_i}{dt^2}$$

Now let's focus on the expression on the right in the parenthesis. Going back to our finite elements on the string. We can use the finite values of y to find the first derivative with respect to x. One way to do this for the element y_i is to find the slope from y_{i-1} (the element before) to y_{i+1} (the element after) to get the following:

$$\frac{dy_i}{dx} = \frac{y_{i+1} - y_{i-1}}{2\Delta x}$$

Note, that this has a change in x of $2\Delta x$ since we are going from the $i - 1$ element to the $i + 1$. The second derivative can be determined by looking at the change in the first derivative. For simplicity, we can look at the change in the first derivative from the midpoint between $i - 1$ and i and then from i to $i + 1$.

$$\frac{d^2 y_i}{dx^2} = \frac{\frac{dy_{i+1/2}}{dx} - \frac{dy_{i-1/2}}{dx}}{\Delta x}$$

Using the same method for finite derivatives gives the following:

$$\frac{d^2 y_i}{dx^2} = \frac{\frac{y_{i+1} - y_i}{\Delta x} - \frac{y_i - y_{i-1}}{\Delta x}}{\Delta x} = \left(\frac{y_{i-1} + y_{i+1} - 2y_i}{\Delta x^2}\right)$$

This finite difference version of the second derivative is not only useful here, but we will bring this back later for numerical calculations using python. However, we can see that in our version of Newton's second law (above) we can replace this finite difference with the second derivative. Also, for the space derivative we are only seeing how the y-value changes with position, this would actually be a partial derivative. The same would be true for the time derivative. With that, we get the following expression.

$$\frac{T}{\mu}\frac{\partial^2 y}{\partial x^2} = \frac{\partial^2 y}{\partial t^2}$$

And this is the wave equation.

$$\frac{\partial^2 y}{\partial t^2} = v^2 \frac{\partial^2 y}{\partial x^2}$$

where the velocity would be:

$$v = \sqrt{\frac{T}{\mu}}$$

2.1   A Solution to the Wave Equation

It's not trivial to create a generic solution to the wave equation for a 1 dimensional string. However, it is possible to pick a function and show that it is a solution. Suppose we have the following function:

$$y(x, t) = A\cos(kx - \omega t)$$

If we want to see if this is a solution, we need to start taking derivatives. Let's start with the space derivatives. The first partial derivative of y with respect to x would be:

$$\frac{\partial y}{\partial x} = -Ak \sin(kx - \omega t)$$

And the second partial derivative:

$$\frac{\partial^2 y}{\partial x^2} = -Ak^2 \cos(kx - \omega t)$$

Now let's look at the time derivatives. The first partial with respect to time:

$$\frac{\partial y}{\partial t} = A\omega \sin(kx - \omega t)$$

And the second partial derivative:

$$\frac{\partial^2 y}{\partial t^2} = -A\omega^2 \cos(kx - \omega t)$$

Putting these two second derivatives into the wave equation:

$$-A\omega^2 \cos(kx - \omega t) = v^2\left(-Ak^2 \cos(kx - \omega t)\right)$$

So, this function is a solution if:

$$v^2 = \frac{\omega^2}{k^2}$$

where k is the wave number and ω is the angular frequency:

$$k = \frac{2\pi}{\lambda} \qquad \omega = 2\pi f$$

With that, we can see that this solution is actually a traveling sinusoidal wave with a wavelength of λ and a frequency f.

2.2   Graphing the Solution

This will be a useful practice in python. Let's create a graph of this sinusoidal traveling wave solution. In order to make a graph, we are going to need numerical values for all the parameters. We can use a string length (L) of 1 m with a mass (m) of 0.5 kg. The tension (T) will be 0.49 Newtons. For the wave, the amplitude (A) will be 0.1 m with an angular frequency of ω = 44 radians per second.

Since this must satisfy the wave equation, we can first calculate the velocity (v).

$$v = \sqrt{\frac{T}{\mu}} = \sqrt{\frac{TL}{m}}$$

With this velocity and the angular frequency, we can then find the wavelength and wave number.

$$v = \frac{1}{2\pi}\omega$$

$$\lambda = \frac{2\pi v}{\omega}$$

$$k = \frac{2\pi}{\lambda} = \frac{\omega}{v}$$

We are going to need a couple more things to create this graph. There will be a variable for time (t) but also the time step (dt). Likewise, we need a position value (x) along with a space step (dx).

Here are these values in python along with the setup for the graph.

```
 1  Web VPython 3.2
 2  g1 = graph(title="It's a Wave", xtitle="x [m]",
 3  ytitle="y [m]",width=500, height=200)
 4  f1 = gcurve(color=color.blue)
 5  L = 1
 6  m = 0.5
 7  T = 0.05*9.8
 8  w = 44
 9  A = 0.1
10  v = sqrt(T*L/m)
11  k = w/v
12  t = 0
13  dt = 0.001
14  x = 0
15  dx = 0.001
16
```

This graph is going to be different than our previous graph. Notice that the function depends on both time and position. Let's use this as an opportunity to create a python function to evaluate the y-position of each part of the string. This python function will act just like the mathematical function in that we will pass it the values for x and y and it will produce the value of y(x, t).

In python, you can create a function with the "def" statement. Here is how that works.

```
17  def y(xx,tt):
18      yt = A*cos(k*xx-w*tt)
19      return(yt)
20
```

- This function has the name of "y" and I'm going to pass to it two variables. In order to avoid conflict with global variables, I'm going to call these xx and tt (instead of x and t).
- Just for clarity, line 18 calculates the value of y for the given values of xx and tt.
- Finally, the function returns the calculated value.

Next we need to think about loops. In our previous graph (mass oscillating on a spring) we only needed to worry about the position of the mass as it changed with time. However, for this string we have to think about the vertical position of every part of the string at different times. Our function for y changes with both time (t) and position (x). One way to deal with this is with a nested loop. We can pick a value of time and then go through all the values of x on the string to calculate our values of y. After that, we can increase the time value and then repeat the whole process.

Here is a nested loop without the graphing.

```
21  while t<1:
22       x = 0
23       while x<2:
24              print(y(x,t))
25              x = x + dx
26       t = t + dt
27
```

Notice that the first loop is over time. A second loop is indented to be part of this first loop and is over space. At the beginning of the time loop (in line 22 above), the x value is reset to zero. Then the space loop (line 23) goes through all the values of x and prints the y position (don't actually run this code). After that, the time value is updated (line 26) and the whole thing starts again.

We don't want to actually print the value of y, we want to make a graph. But how do we make a graph for a variable (y) that changes with both position (x) and time (t)? One way is with an animated graph. Yes, I know this is a static-image book so that you can't see an animated graph. However, I'm going to show you how to make it anyway.

In Web VPython, we can make an animated graph by first making a graph object (just like our previous graph). We also need to create a gcurve object to plot points on our graph (just like before).

To make an animated graph, we are going to use the following steps. First, instead of plotting each point using f1.plot (t,y) we will add points to a list of data points. After we have this full list of y versus. x values, we can then plot it. After that, we will delete the list of data points and move forward in time to create the list for the next time step to plot.

We can control the speed of the animated graph using the rate(100) statement. This will be placed in the time-loop and it tells python to not run more than 100 loops every second. If you change it to rate(200) it should run the animation faster—assuming that the code is simple enough that your computer can keep up.

Here is the code with the loops and the animated graph.

```
21  while t<1:
22       rate(100)
23       x = 0
24       f1data = []
25       while x<2:
26              f1data = f1data + [[x,y(x,t)]]
27              x = x + dx
28       f1.data = f1data
29       t = t + dt
30
```

The output would look like this:

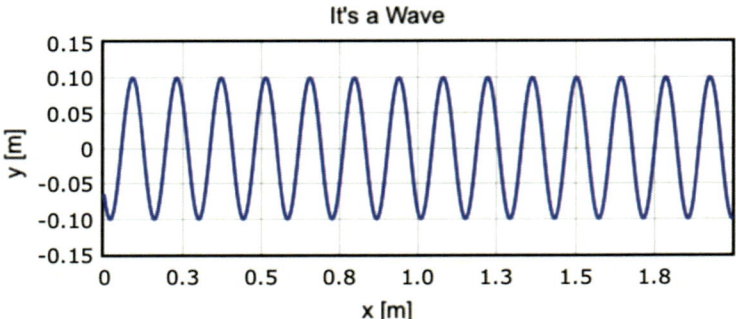

Except that the curve would be moving to the right. You will just have to either run this code on your own or use your imagination.

Modeling a Wave with Many Masses and Springs

In the previous chapter, we explored a wave on a string by imagining it to be split into finite elements of mass separated by massless sticks. From that, we were able to allow the pieces to get infinitely small and become derivatives. This gave us the wave equation for a string.

But what if we didn't let the pieces get infinitely small? What if they were just small? Suppose we have a string that's represented by just 10 masses connected by springs? Would we be able still get the same motion for a wave on this string? The answer is of course, yes.

Here's what we are going to do—we are going to build a python model consisting of a finite number of masses connected by springs. There are a lot of pieces to this model, so we need to start small.

3.1   Two Masses and One Spring

The simplest model of a string would be just two masses with a spring connecting them. It's actually too simple to be useful to explore waves, but it's going to be helpful to set up a more complicated python program. Again, the main idea is that we will break the motion into short time intervals. During each time interval we can assume the net force is constant to determine the change in momentum of each object. Here's the basic step for each time interval.

- Use the position of each mass to calculate the stretch of the spring connecting the two masses.

© The Author(s), under exclusive license to Springer Nature Switzerland AG 2025
R. Allain, *Modeling Waves with Numerical Calculations Using Python*, Synthesis
Lectures on Wave Phenomena in the Physical Sciences,
https://doi.org/10.1007/978-3-031-78291-6_3

- Calculate the force (from the spring) on each mass.
- Use the net force to update the momentum of each mass.
- Use the momentum to update the position of each mass.
- Repeat.

We've already seen the two objects in Web VPython that we are going to need (the sphere and the helix), so let's get started. Of course, this is a numerical calculation, that means that we are going to need some numbers for the initial positions and momentums of the two masses and the spring constant.

Here's the start of the code. Comments to follow.

```
 1 Web VPython 3.2
 2 canvas(background=color.white)
 3 m = 0.01
 4 k = 1
 5 L0 = .05
 6 t = 0
 7 dt = 0.01
 8
 9 m1 = sphere(pos=vector(-L0/2,0,0), radius=L0/20,color=vector(.5,.5,.5))
10 m2 = sphere(pos=vector(L0/2,0,0), radius=L0/20,color=vector(.5,.5,.5))
11 spring = helix(pos=m1.pos, axis=m2.pos-m1.pos, radius=L0/30,
12 thickness=0.0005,coils=10)
13 m1.m = m
14 m2.m = m
15 m1.p = m1.m*vector(0,0,0)
16 m2.p = m2.m*vector(0,0,0)
17
18
```

Comments:

- Lines 3–7 are just our constants. Yes, these values have units but I left them out to avoid clutter (they would just be in there as a comment).
- In line 5, L0 is the unstretched length of the spring.
- The masses are m1 and m2 and they are 3D spheres. I added a color vector so that they show up better against a white background.
- Lines 13 and 14 set the mass values as a property of the m1 and m2 objects. It might look weird, but this is a very useful way to code to keep track of all the different constants (especially when we have many more objects). In this code, the two masses are at the equilibrium position for the spring.
- Lines 15 and 16 set the initial momentums for both masses (in this code they start off at rest).

Here's what it looks like.

Now let's make it move. Just like before, we can make a loop over the time intervals. In order to make it interesting, I'm going to start the left mass (m1) displaced a small amount to the left. Here's the rest of the code.

```
18  while t<3:
19      rate(100)
20      r12 = m2.pos-m1.pos
21      F12 = -k*(L0-mag(r12))*norm(r12)
22      m1.p = m1.p + F12*dt
23      m2.p = m2.p - F12*dt
24      m1.pos = m1.pos + m1.p*dt/m1.m
25      m2.pos = m2.pos + m2.p*dt/m2.m
26      spring.pos = m1.pos
27      spring.axis = m2.pos-m1.pos
28      t = t + dt
29
```

Comments.

- Line 20: We are going to have to do things a little different than before. First, we need a vector pointing from mass 1 to mass 2. This will be used to determine the stretch or compression of the spring.
- Line 21: This is the force calculation just like we did previously for the single oscillating mass. Notice that I'm using the notation "12" to mean the force that 1 exerts on mass 2.
- Lines 22–23: Updating the momentum of the two masses. Notice that for mass 2, we are using negative F12. From Newton's third law, the force that mass 2 exerts on mass 1 is the opposite of the force 1 exerts on 2.

Just for completeness, let's plot the x-position of the masses as a function of time.

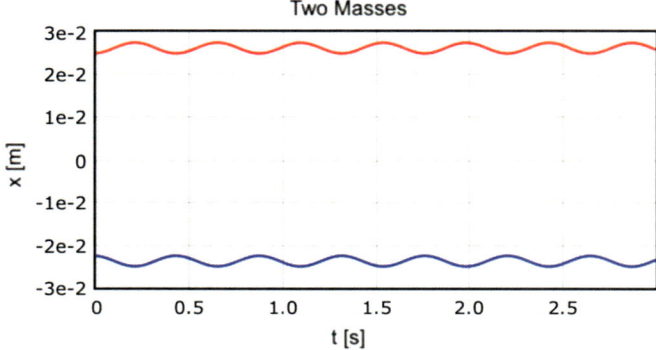

Note that it's possible to give each mass an initial momentum in the y-direction, but that's just going to make the system rotate (which is cool, but not really much like a wave on a string).

3.2   Three Masses

If we can do two masses, we should be able to do three masses with two springs. Things are going to get a little more complicated, so let's start with a diagram.

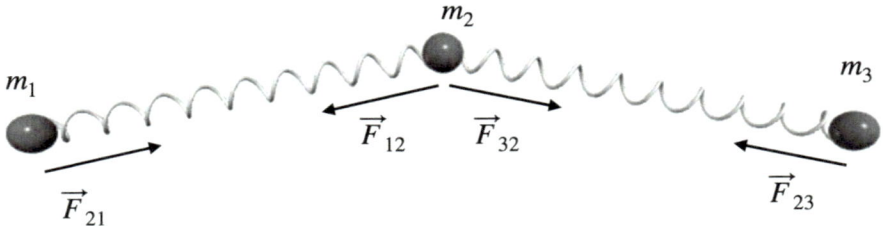

It's important to have some consistent rules for labeling these three objects. We can then use the same naming conventions for many masses. Also, you can see that there are four forces, but we really only need to calculate 2 of them. In the previous code, we calculated F12—so we will keep that. Then we can calculate F23 (always going forward in the list of masses).

Here is the new setup.

```
12  m1 = sphere(pos=vector(-L0,0,0), radius=L0/20,color=vector(.5,.5,.5))
13  m2 = sphere(pos=vector(0,L0/5,0), radius=L0/20, color=vector(.5,.5,.5))
14  m3 = sphere(pos=vector(L0,0,0), radius=L0/20,color=vector(.5,.5,.5))
15  spring12 = helix(pos=m1.pos, axis=m2.pos-m1.pos, radius=L0/30,
16  thickness=0.0005,coils=10)
17  spring23 = helix(pos=m2.pos, axis=m3.pos-m2.pos, radius=L0/30,
18  thickness=0.0005, coils=10)
19  m1.m = m
20  m2.m = m
21  m3.m = m
22  m1.p = m1.m*vector(0,0,0)
23  m2.p = m2.m*vector(0,0,0)
24  m3.p = m3.m*vector(0,0,0)
25
```

There's really not that much new here. Let's jump to the time loop.

```
27  while t<3:
28      rate(100)
29      r12 = m2.pos-m1.pos
30      r23 = m3.pos-m2.pos
31      F12 = -k*(mag(r12)-L0)*norm(r12)
32      F23 = -k*(mag(r23)-L0)*norm(r23)
33      m1.F = -F12
34      m2.F = F12 - F23
35      m3.F = F23
36      m1.p = m1.p + m1.F*dt
37      m2.p = m2.p + m2.F*dt
38      m3.p = m3.p + m3.F*dt
39      m1.pos = m1.pos + m1.p*dt/m1.m
40      m2.pos = m2.pos + m2.p*dt/m2.m
41      m3.pos = m3.pos + m3.p*dt/m3.m
42      spring12.pos = m1.pos
43      spring12.axis = m2.pos-m1.pos
44      spring23.pos = m2.pos
45      spring23.axis = m3.pos-m2.pos
46      t = t + dt
47
```

There's not too much different in this three mass model. However, there are some important points.

- For the net force for each object, I'm using that as a property of the object. That means that m2.F is the net force for mass 2. This is useful when dealing with more than one mass—but it's going to be critical when creating a string with many masses.
- Notice that m1 and m2 are at the "end of the string" and they are allowed to move up and down. For a model of a more realistic string, we might choose to make the end points fixed.
- I'm trying to use consistent notation for the calculations–always going from one mass to the next for the r-values and then using those vectors to calculate the forces.

It looks like this model works fine–for three masses. But it's not really much like a string. It's indeed possible to use this exact same method for more masses. However, once you get more than 5 masses it just gets very tedious to keep track of all the objects. We are going to need some better python tools to effectively model a string with more masses.

3.3   Python Lists

Imagine we want to make a string with 10 masses. If we keep the end masses in a fixed position (like a string with the ends stationary) then you only need 8 masses. However, connecting each mass to the next means you will need 9 springs. That means that the python code will get messy very quickly. In order to get around this problem we are going to have to use a python list.

In python, a list is a variable that can contain more than one thing. In a sense, you can think of a vector as a list of three variable (x, y, and z)–but the way we use a vector is a better than just a list. Maybe you could think of a list as a grocery list. It's pretty much just like that.

OK, let's just make a list.

```
1  Web VPython 3.2
2
3  stuff = [1, 3.3, 4.4, -2.1]
4
5  print(stuff)
6
7
```

It's just a list of numbers. When you run this, it prints the following.

```
[1, 3.3, 4.4, -2.1]
```

No surprises there. Just a couple of comments about lists:

- You can give the name of a list just like you would for any other python variable (the name can't start with a number).
- The list is created by putting items in between square brackets "[]" with each element separated by a comma.
- The list items can pretty much be ANYTHING. They don't have to be numbers. A list can even be a mix of numbers and strings (but we won't do that).
- You can even make a list with vectors or a list containing other lists.

But a list by itself isn't really that useful. We still need to address individual items in the list. Suppose I want to print the second number in the list above. I can do that with the following.

```
4
5  print(stuff[1])
6
```

Yes, each element has an index. In this list example, there are 4 items and they have an index starting from the left of 0, 1, 2, 3 with 0 being the first element and 3 being the last.

Here are some other useful features of lists and their indexes.

- If you have a list and you don't know the length, you can find it with len(stuff). That returns a number equal to the number of items in the list.
- It's possible to index an item backwards. If you use stuff[−1] that will address the last item in the list. You can continue with negative values such that stuff[−2] is the second to last item.
- You can change an individual item in a list. Suppose you want to change the second item in stuff to the value of 0.12, you would just use stuff[1] = 0.12. Done.

OK, let's do some useful stuff with lists for practice. Let's create a list of 10 random numbers. Just a quick note—in Web VPython, there is a simple random number generator "random()". This returns a random number between 0 and 1. In order to make a list of random numbers, we can use the following plan.

- Make a loop that counts up to 10. There are several options to count like this, but I'm going to use a while loop (since that's what we use before).
- In the loop, create a random number.
- Add this number to the list.
- Repeat.

Here's what that code looks like.

```
 3  N = 10
 4  rando = []
 5  n = 1
 6
 7  while n<=N:
 8      temp = random()
 9      rando = rando + [temp]
10      n = n + 1
11
12  print("random numbers = ", rando)
13
```

In this case, I'm using N for the maximum number of items with n as the "counter". Notice that the while loop needs to run as long as n is less than OR EQUAL to N. This is because n is incremented in line 10. You want to run through the loop one more time after n gets up to 10. Don't worry about this though. Often times we will deal with large values of N and there's not much difference between a list that is 1000 elements long and one that's 999 long. But if you are concerned, you can always just play with the code and print the length of your final list.

There are two more important parts of this code. First, look at line 4. Here we are creating a list, but it's empty (just []). You can add anything to a list that doesn't exist, so we need to make something first. The second important part is in line 9. We can add an element to the list by just adding [temp] to the current list. This makes it look just like the momentum update formula that we have used in the past. It's possible to use the "append" function instead. You could alternatively replace line 9 with "rando.append(temp)".

We can add to a list, but is it also possible to do other stuff to a list? Yes, try the following commands.

- rando.pop(2). This removes item with an index of [2] (the third item) from the list.
- rando.insert(2,3). This puts the number 3 at the [2] position.
- rando.clear(). This removes everything from the list.

Now for something very important when dealing with lists–traversing a list. The best way to learn is by doing–so let's make a list of vectors and then go through the list and print the magnitude of each vector.

The first part is to create a list of vectors. Imagine that this is a list of position vectors ($\vec{r}$).

```
3  rs = [vector(1,2,3), vector(-1, 0, -1), vector(2, -1, 1),
4            vector(-3,-2,0), vector(1,1,1)]
5
```

One way to go through the list of items is with the for loop. It looks like this.

```
6  for r in rs:
7      print(mag(r))
8
```

This is actually pretty awesome. It goes through the list rs and gives each item a temporary name of r. In the loop, you can do whatever you like to the variable r and it just changes and moves to the next element. In this case, we just take the magnitude of the vector and print it. Of course we will do more useful things later.

Just for completeness, here's another way to do the exact same thing.

```
6  for n in range(len(rs)):
7      print(mag(rs[n]))
8
```

Although this code outputs the same result, it is indeed different. This again uses the for loop, but it has a variable n going from 0 to the length of the list rs. Then to access the elements in the list, we use the rs[n] index. It looks more complicated, but this method is also very useful. Imagine that we have two lists of vectors, r1 and r2 and we want to find the magnitude of the difference between these vectors. That code could look like this (assuming both lists are the same length).

```
 8  for n in range(len(r1)):
 9      s = r2[n] - r1[n]
10      print(mag(s))
11
```

Using this index loop, we can go through both lists at the same time.

Now let's make another list. Suppose I want to make 10 balls in space moving with random velocities. Yes, it's not a wave on a string, but it's going to be very useful. However, you will see that we can make a list of sphere objects and it will make our code much simpler.

Let me just jump to the code and then we can go over the important parts.

```
 1  Web VPython 3.2
 2
 3  #length of box
 4  L = 1
 5  #number of balls
 6  N = 10
 7  #max velocity
 8  v0 = 0.2
 9
10  balls =[]
11  n = 0
12  #make balls
13  while n<=N:
14      tempr = vector(0.5*L*(2*random()-1),0.5*L*(2*random()-1),0.5*L*(2*random()-1))
15      balls = balls + [sphere(pos=tempr, radius=L/20, color=color.yellow)]
16      n = n + 1
17
18  #set initial velocity
19  for b in balls:
20      b.v = v0*vector(2*random()-1,2*random()-1,2*random()-1)
21
22  t = 0
23  dt = 0.01
24
25  while t<1:
26      rate(100)
27      for b in balls:
28          b.pos = b.pos + b.v*dt
29      t = t + dt
```

There's a bunch of stuff here, let me start with the list of balls. Notice that I again go through a loop N times. For each run through the loop, I create a random vector between − L/2 and L/2 in each direction. This can be accomplished with (L/2) (2*random()−1). Remember that random() is a number between 0 and 1 so that 2*random()−1 is between − 1 and 1. Then this is multiplied by L/2.

Next, I use this random vector as the position of a sphere that is added to the balls list–yes, it's a list of 3D sphere objects. Of course these balls also need an initial velocity.

In line 19, I go through the list of balls and assign a velocity property to each object as a random vector.

Finally, I need to animate each ball. This last loop is a time-based loop like we did with the momentum principle before. The balls are moving with a constant velocity, so I don't need to worry about forces or change in the momentum. Instead, I just have to update the position of the balls–each ball in the balls list. This is the perfect place for a simple list traverse that you see in lines 27–28. It goes through each item in the list of balls and updates its position.

In the end, you get something like this.

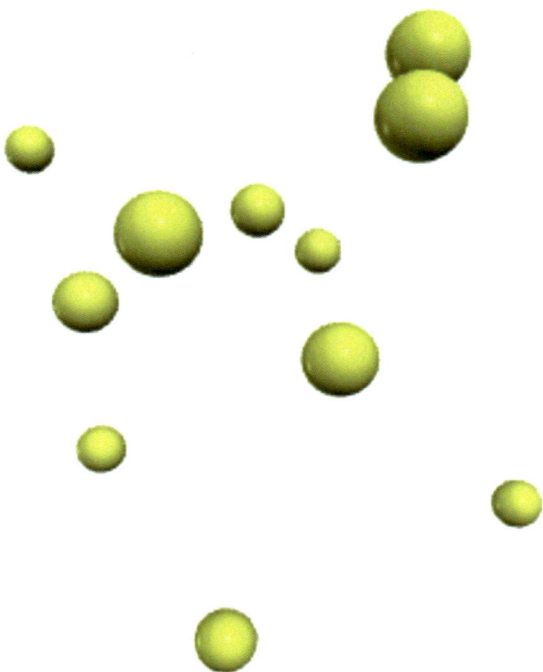

Except the balls move. But here's the awesome part. If you want to create 1000 balls instead, you just have to change the value of N to 1000. That's it. It's also possible to make a bunch of balls that collide with the walls of a container or even interact with other balls, but that's not super useful for a wave on a string. The important thing is that you are a little more familiar with lists in python.

3.4   A String Made of Many Balls

Now that we have a method to model two balls connected by a spring as well as a nice method for handling many balls (using python lists) we can put these two things together and model a wave on a string in python.

   We have already created a python model with 3 masses, we just want to extend that to many masses. With that, we have to realize that most of these masses will have two springs connected to them with two forces. However, the two masses on the end will only have one spring force on them. Also, just to make this act like a fixed string, these two end masses will be fixed in place.

   There is one additional thing to think about before building this model. Imagine that we have a real string with a length, a mass, and a tension. These macroscopic properties of the string are important so that we can use the wave equation that we derived earlier. But if we break the string into finite elements then we also need the properties for each individual mass and for the tiny springs. You can think of these as the microscopic properties of the string. There needs to be a connection between the macro and microscopic properties.

   Let's say that we have a string with a length of $L = 1.2$ m and a total mass of 0.025 kg. If this string has some stretch to it, there will also be a macro spring constant. We can call this K and set it equal to 1.64 N/m (these values were chosen just because things work fine with them).

   Now for the macroscope properties. If we use 34 masses ($N = 34$) then each of these masses will be 0.025/N kg. Also, the distance between these masses will be $L/(N - 1)$. Finally, these masses are going to be connected by small springs. In order to have the macroscopic spring constant related to the micro spring constant, we can think of a bunch of tiny springs in series. Suppose you have two springs in series. If you pull with a force F on these springs, they will stretch some distance s. However, since BOTH springs will stretch the individual springs need to have a constant that's twice the value of the two in series. We can expand this to a bunch of springs such that the individual spring constants would be $k = K*(N - 1)$ so that this will match the overall string properties.

   The next step is to build our list of balls and springs to model our wave. We are going to need some tricks though. As masses are added to the list of balls, we need to start with some vector location. I'm going to put this starting location at $-L/2$ on the x-axis. The other end of the string will then be that starting location plus a vector of length L in the x-direction. Finally, I just need the "step" size. This will be the total length of the string divided by $L/(N - 1)$.

   Here is the code to set up the balls and the springs.

```
 5  L=1.2
 6  M=.025
 7  K=1.64 #total spring constant (in theory)
 8
 9  N=34 #number of masses
10  k=K*(N-1)
11
12  leftend=vector(-L/2,0,0)
13  ds = vector(1,0,0)*L/(N-1)
14  R = L/(5*N) #ball radius
15
16  balls = []
17
18  for i in range(N):
19      balls = balls +[sphere(pos=leftend+i*ds,radius=R, m=M/N,
20      p = vector(0,0,0),F = vector(0,0,0))]
21
22  springs = []
23  for i in range(N-1):
24      springs = springs + [cylinder(pos=leftend+i*ds, axis=ds, radius=R/2)]
25
26  L0 = 0.9*L/(N-1)
27
```

Notice that there are two lists of objects, one for the balls and one for the springs. In the case of the balls, I have a loop (line 18) that counts up to N using the counter variable i. This counter is then used to move the position of each starting mass a particular distance.

There are some other important properties for each ball in the string. Along with the position, I also create a value for the mass (m), the initial momentum (p) and the net force on that mass (F). There is one fewer springs than balls so that the counting loop goes over N − 1 items.

One last thing. In order to calculate the spring force from each spring, I need to also know the value of the unstretched length–which I am calling L0. This is calculated in line 26. With that, we get the following output.

It doesn't matter that this is a static image–nothing happens anyway (because we aren't finished). It looks nice though. Now for some motion. Of course no matter what code we add to this program, nothing will happen. The masses are in an equilibrium position so that the net force is zero. They start at rest and they will stay at rest. Just to get things started, let's give a tiny displacement to just one of the masses. I'm going to add the following.

```
32  balls[5].pos=balls[5].pos + vector(0,L/40,0)
33
```

This moves the vertical position of the 6th mass up in the y direction by a value of L/40. With that one mass out of equilibrium, it will have a non-zero net force and change it's momentum. This will exert forces on the other masses and the whole string should oscillate.

Here is the rest of the code.

```
30  t = 0
31  dt = 0.0001
32
33  while t<.32:
34      rate(1000)
35
36      for i in range(1,N-1):
37          balls[i].F =( -k*(mag(springs[i-1].axis)-L0)*norm(springs[i-1].axis) +
38          k*(mag(springs[i].axis)-L0)*norm(springs[i].axis))
39
40      for ball in balls:
41          ball.p = ball.p + ball.F*dt
42          ball.pos = ball.pos + ball.p*dt/ball.m
43
44      for i in range(1,N):
45          springs[i-1].axis = balls[i].pos - balls[i-1].pos
46          springs[i-1].pos = balls[i-1].pos
47
48      t = t + dt
49
```

There's some stuff that looks fairly complicated here–but just keep in mind that it's really not any different than our three mass string from before. Just a quick reminder, we need to set values for the initial time (t = 0) and the time step (dt = 0.0001). Yes, this time step is rather small since the masses are close together. There is also the rate (1000) statement in line 34. This tells the code to run 1000 loops per second (maximum rate). But the time steps are 1/10,000 s. This means that the animation will run in "slow motion" so that we can see what's going on.

In line 36 we go through most of the balls in the list to calculate the net force. Notice that I am using the "range" method to traverse the list of balls. This is important because now I can use the code "for i in range(1, N − 1):". This counts the variable i starting with a value of 1 and going up to N − 1. That means that balls[i] will never get to the first or last ball. That's nice since those two masses are special in that they don't move and they only have one spring connected to them.

For the actually force calculation, it looks complicated but it's identical to our 3 ball string. There are two force on each mass. There's a force from the mass on the "left" which would be [i − 1] and then one from the right side. Notice that I'm using the vector length of the spring from the springs[] list. All the springs attach from the left mass to the right that means that for the right side force I'm just using spring[i].axis (the vector length of the spring). These lengths are updated later. Also, it should be noted that line 37 is wrapped around to line 38 so that it would fit in this text format.

The next loop starts on line 40. This goes through all of the balls in the balls list and updates both the momentum and the position. Here is an important point. The two end balls do indeed have a net force of <0, 0, 0> from the initial setup. The force calculation for the springs never changes these two net force values so they are still at zero and the momentums and positions of these two masses never change (fixed ends of the string).

Finally, we have one more loop starting at line 44. This goes through the list of all the springs and resets both the position and the axis (the vector length of the spring). This axis value is important for the force calculation.

If you run this code, it does indeed work, but it's not very exciting. It's just a simple string that is shaking around. It doesn't look like a wave. But does it act like a wave? Let's see if we can make a connection between the wave speed from the wave equation and the speed of a pulse on this ball string.

Recall that the a wave will travel along the string with a velocity that depends on the tension and the linear mass density.

$$v = \sqrt{\frac{T}{\mu}}$$

The linear mass density is fairly easy to calculate since we already have the total mass and the length of the string. What about the tension? The way that our mass-ball string was created in equilibrium, the string tension is actually zero newtons. We can fix this by defining the unstretched length (L0) as something smaller than the distance between masses.

47 L0 = 0.9*L/(N-1)

Now we can calculate the tension in the whole string just by looking at the spring force in one of the springs (it doesn't really matter which one). Using the values here, I have a string tension of 0.1968 Newtons. The linear mass density is 0.0208 kg/m. This gives a theoretical wave speed of 3.073 m/s.

The next step is to create a wave pulse on the string. We can do this by oscillating the mass on the left end up and down with a y-component given by:

$$y(t) = A \sin(\omega t)$$

where A is the amplitude of the pulse and ω is the angular frequency. We can give the pulse an amplitude of 0.03 m with an angular frequency of 45 radians per second. If we want just half a full wave pulse then this mass will only oscillated until the time gets up to $\frac{\pi}{\omega}$.

Here's a snapshot of our new wave pulse.

But how fast is this pulse moving? We have a problem. How do you measure the horizontal speed of a vertical displacement? Let's make a trick to get this calculation. I

want to look at my list of balls and find the one ball that has the largest y-value. This will
give us the x-position of the peak. Here is a function that does just that.

```
4  def maxxy(stuff):
5      ymax=0
6      xmax=0
7      for bb in stuff:
8          if bb.pos.y>ymax:
9              ymax = bb.pos.y
10             xmax = bb.pos.x
11     return(xmax)
12
```

This python function takes a list of objects (the balls) and returns the x-value of the
ball with the largest y-value. Now as the pulse moves along the string, we can create a
plot of the horizontal position of the pulse as a function of time.

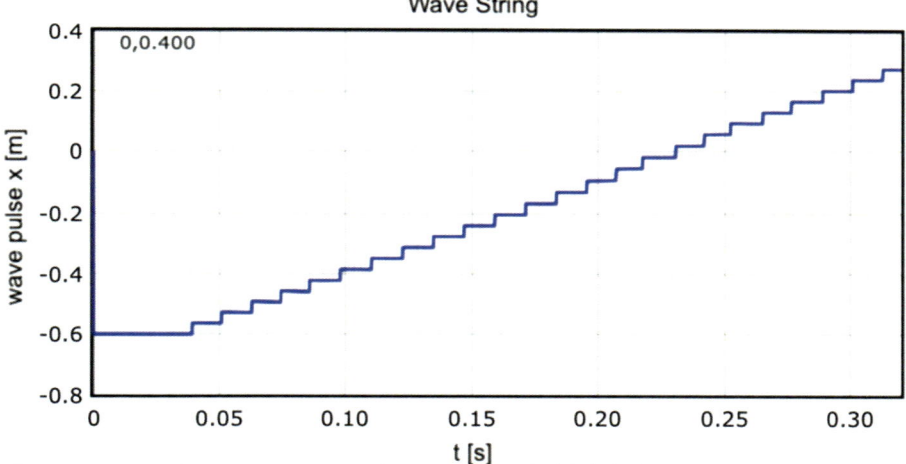

The slope of this line will be the wave velocity. With some simple calculations, I get
a velocity of 3.2 m/s. It's not quite the same as the theoretical speed but this isn't quite a
theoretical wave on a string since it is created from only 34 masses.

Oh, but why isn't the position time graph smooth? Since we have a finite number
of elements, it's possible that a single mass can stay in the position of greatest y-value
for some short time interval. With the masses only moving up and down, there are only
discrete values that are possible for the x-position. This creates that "stair step" plot of
position.

Wave on a String with the Finite Difference Method

Of course there is another method to model the motion of a wave on a string without having to essentially build a bunch of masses and springs (even if we only built those in python and not in real life).

Let's again consider our string as a bunch of finite masses connected together. However, instead of explicitly calculating the force on each mass and using that to update the momentum and position of the masses, let's start with the wave equation.

$$\frac{d^2y}{dt^2} = \frac{T}{\mu}\frac{d^2y}{dx^2}$$

This is the one dimensional form of the wave equation with masses moving up and down in the y-direction as the wave moves in the x-direction.

If we think about the masses at discrete locations and at discrete times, then each mass can have a position index (I will use i) and a time index (using k).

$$y_{i,k}$$

We can find the first derivative with respect to position at point i by looking at the point before (at $i - 1$) and after (at $i + 1$). The rate of change from these two points would be:

$$\frac{dy_{i,k}}{dx} = \frac{y_{i+1,k} - y_{i-1,k}}{2\Delta x}$$

© The Author(s), under exclusive license to Springer Nature Switzerland AG 2025
R. Allain, *Modeling Waves with Numerical Calculations Using Python*, Synthesis
Lectures on Wave Phenomena in the Physical Sciences,
https://doi.org/10.1007/978-3-031-78291-6_4

where Δx is the horizontal spacing between points. Notice that we STILL have a time index (k) but that doesn't change with the x-derivative. So, technically, this would be a partial derivative with respect to x.

Just like we did previously, we can use the derivative at the halfway points between i − 1 and i as well as the halfway between i and i + 1 to get the second derivative (the second partial).

$$\frac{d^2 y_{i,k}}{dx^2} = \frac{y_{i+1,k} + y_{i-1,k} - 2y_{i,k}}{\Delta x^2}$$

If we want to model the wave equation, we need to also consider a partial derivative with respect to time. The same finite different process can be repeated with the time index (k) to get the following expression.

$$\frac{d^2 y}{dt^2} = \frac{y_{i,k+1} + y_{i,k-1} - 2y_{i,k}}{\Delta t^2}$$

Here the Δt is the finite difference between successive times. It's same as the time step that we have used in the previous numerical calculations.

Using these two definitions for the second derivatives, we can rewrite the wave equation as the following.

$$\frac{y_{i,k+1} + y_{i,k-1} - 2y_{i,k}}{\Delta t^2} = \frac{T}{\mu}\left(\frac{y_{i+1,k} + y_{i-1,k} - 2y_{i,k}}{\Delta x^2}\right)$$

There's a bunch of stuff going on here–so, let's be clear. For any term that is an "i", that is the current element in space. The i + 1 terms are items to the right (in the positive x-direction) and i − 1 are to the left. For the time index, "k" means "now". The k + 1 terms are the future and k − 1 is the past.

Finally, since Δx and Δt are constant along with T and μ we can group all of these together and call it the constant R.

$$y_{i,k+1} + y_{i,k-1} - 2y_{i,k} = R\left(y_{i+1,k} + y_{i-1,k} - 2y_{i,k}\right)$$

$$R = \frac{T\Delta t^2}{\mu\Delta x^2}$$

We can solve this expression for the value of y at the position element i in the future (time index k).

$$y_{i,k+1} = R\left(y_{i+1,k} + y_{i-1,k} - 2y_{i,k}\right) + 2y_{i,k} - y_{i,k-1}$$

Notice that the only instance of the future value for y is on the left side of the equation. On the right side, the only time elements are either now (k) or in the past (k − 1).

However, for the space elements we need both the values on the left (i − 1) and right (i + 1) of the current element.

We can use this expression to find the future values of y for all the elements in our string. Of course, there are some limitations. In order to solve for the future (k + 1) values for y, we need to know both the current values (k) and the values before that (k − 1). This means that for the start of the calculation we need two previous iterations to begin with. There are a couple of ways to deal with this, but you need to know that it's an issue.

In terms of space, if I want to solve for the y value at the ith position, I need the values to the left (i − 1) and right (i + 1). This is only a problem at the endpoints. The element on the far left (i = 0) doesn't have an element to the left–there isn't a value for the i = − 1 position. The same is true for the last element on the right. Actually, this isn't a problem if we keep the endpoints fixed. In that case we don't need to update the positions for i = 0 or i = N since they don't change.

Here is the plan.

- Break the string into finite elements. Each of these elements will have a vertical position (y).
- These element values will be stored in 3 separate lists in python. There will be a "now" list for the k time element. There will be an "old" list for k − 1 elements and then there will be a "new" list for k + 1.
- For the initial conditions, I will set the values of both the "now" and "old" list to some particular values.
- With these lists, I can go through the position values of the "old" and "now" lists to calculate the values for the future ("new") list.
- Once the new values are calculated, the now list will be changed to "new" and the old list to "now".

For this model, I'm going to use the following parameters.

- Length: L = 1 m
- Tension: T = 0.04 N
- Linear mass density: μ = 1 kg/m.

It turns out that the choice of both dx and dt can be critical. In this case, I'm going to use a space step size of dx = 0.01 m and a time step of dt = 0.05 s.

The first step will be to create some lists. I'm going to actually need four lists. There will be a list of x values for the locations of the mass elements in the string (separated by distance dx). Then I will need three lists for the y positions of these mass elements. I'm going to call these y values: yold, ynow, ynew. Here's the code for the constants.

```
 3 g1 = graph(xtitle="x",ytitle="y",width=400, height=200,
 4 ymax = 0.15, ymin = -0.15)
 5 f1 = gcurve(color=color.blue)
 6 dx = 0.01
 7 dt = 0.05
 8
 9 L = 1
10 T = .04
11 mu = 1
12
13 R = (T/mu)*dt**2/dx**2
14 v = sqrt(T/mu)
```

Here is the code to set up the four lists.

```
17 xt = 0
18 x = []
19 while xt<=L:
20     x = x + [xt]
21     xt = xt + dx
22 x = x + [xt]
23
24 yold = []
25 ynow = []
26 ynew = []
27
28 for i in range(len(x)):
29     yold = yold +[0]
30     ynow = ynow +[0]
31     ynew = ynew +[0]
32     .
```

The first list to create is for the x-values. Here we have a temporary value for x (called xt). We can start this xt value at zero and then increment by an amount dx until it gets to the end of the length of string.

For the three lists with the y-values, it's important that they are the exact same length as the list with the x-values. This means that combining the x and y lists together would give a series of (x, y) data point for the string. To accomplish this, a for loop is create over the range of the x list (that's like 28). After that, zero values are added to the y lists. Of course, if all the y-values are zero nothing will happen with our string. Just to start off, let's pick an element (element number 25 just as a random choice) and set its y-value to 0.01 m.

Now for the change in the string elements over time. This is the code for that.

```
51 t = 0
52 while t <5:
53     rate(50)
54     f1data = []
55     for i in range(1,len(x)-1):
56         ynew[i]=2*ynow[i]-yold[i]+R*(ynow[i+1]+ynow[i-1]-2*ynow[i])
57
58     for i in range(len(x)):
59         yold[i] = ynow[i]
60     for i in range(len(x)):
61         ynow[i] = ynew[i]
62     for i in range(len(x)):
63         f1data = f1data +[[x[i],ynew[i]]]
64     f1.data = f1data
65     t = t + dt
66
```

Of course this is a large loop over time. Using a rate of 50 loops per second, it's going to run in "slow motion" which will allow us to see what's actually going on as it runs.

There are a couple of lines here that are useful to make an animated graph in Web VPython. You obviously won't be able to see these animations in this text, but I'll briefly explain how it works. The curve in this case is named "f1". In a normal graph, you would just plot a data point using f1.plot(x, y)–where x and y are the horizontal and vertical values. However, if we want these values to change over time we need to do something different.

The first step is to create a list of data points that will be plotted. In this instance, that list is called f1data. In line 54 that list is set to an empty list. Notice that this list is set to empty at the beginning of the time-loop. That's important at it clears out the old data for each new value of time.

When new values for the y positions of each element are calculated, an x-y data point is added to this f1data list (see line 63). Notice that the x-y data point is added as a 2 element list to the full data list. Finally, once the f1data list contains all the points, it's plotted (in line 64). Since this list is cleared at the beginning of the next loop, the resulting graph appears to be animated. But again, these details are only for the animation.

For the rest of the time loop, we have the full finite difference calculation. The main part is the loop in lines 55–56. Notice that this loop indexes over the length of the list of x-values but it starts at 1 (instead of 0) and only goes up to 1 less than the length. This means that if we use the index i, we won't change the values for the first or last element.

If you look carefully, you should be able to see that line 56 is identical to the equation above for $y_{i,k+1}$ but here we are calling that element ynew[i] (remember that the "new" versions are the future that correspond to the k + 1 time element).

After the new values of y are calculated, we can update the old and now values. This is what happens in lines 58–59. We don't have to worry about skipping the first and last element since they were zero and stays zero. Also, it's important to update the old values and then update the now values after that (in lines 60-1). Other than updating the value of time, that's it.

When this runs, you might get something like this.

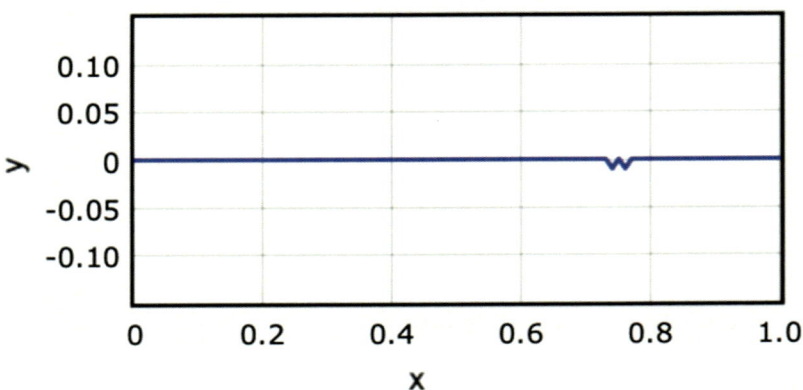

But this is not really want we wanted. Let's create a better wave pulse. How about a wave pulse that's half of a full wavelength with an amplitude of 0.1 m and a width of 0.1 starting at x = 0.2 m. Here's the code for that.

```
34  w = 0.1
35  A = 0.1
36  x0 = 0.2
37  for i in range(len(x)):
38      if x[i]>=x0 and x[i]<=(x0+w):
39          ynow[i] = A*sin(pi*(x[i]-x0)/w)
40
```

The loop in line 37 goes through all the x values. If the current x value is between the start of the pulse (x0) and the end (x0 + w) then the ynow value is set to a sine function with a wavelength of 2 * w. However, this single loop doesn't give us the desired effect. If you run this as it is, then it would be like displacing a small portion of the string with this shape and letting it go. Instead, we want a wave pulse to travel in the positive x-direction with a velocity v. To do that, we need to also set the values for yold.

```
41  xn1 = x0-v*dt
42  for i in range(len(x)):
43      if x[i]>=xn1 and x[i]<=(xn1+w):
44          yold[i] = A*sin(pi*(x[i]-xn1)/w)
45
```

In order to set these yold values, we need to go back in time by a time step of dt. This means the new starting location of the wave pulse is going to be xn1 and it's shifted back by an amount v * dt (where v is the velocity of the wave based on the wave equation).

Now that the first two times for the values of y are set, the pulse is ready to move. This is what it looks like at two different time values.

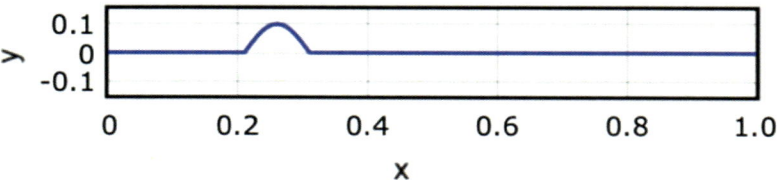

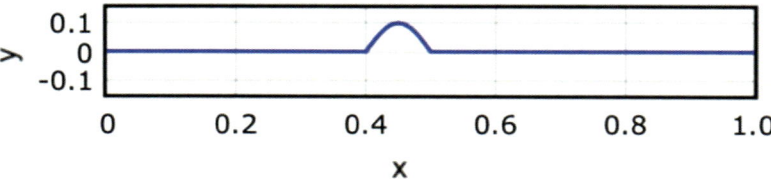

Looks good.

Special Case: Standing Waves on a String

5

Now that we have a couple of methods for solving the motion for a wave on a string, let's apply it to a special case–standing waves. A standing wave happens when the ends of a string are fixed and motion is initialized such that the resulting wave has a wavelength that fits onto the string. This is exactly what happens when a guitar string is plucked.

Let's start off with an analytical solution. Recall that for a one dimensional string in the x-direction, the following wave equation must be satisfied.

$$\frac{\partial^2 y}{\partial t^2} = v^2 \frac{\partial^2 y}{\partial x^2}$$

Where y is a function of x and t and v is the velocity of the wave. We have seen that a sinusoidal function is indeed a solution to this wave equation.

$$y(x,t) = A\sin(kx - \omega t)$$

where k (the wave number) and ω (the angular frequency) must satisfy:

$$\frac{k^2}{\omega^2} = v^2$$

We can define the wave number as:

$$k = \frac{2\pi}{\lambda}$$

where λ is the wavelength and the angular frequency is defined as:

$$\omega = 2\pi f$$

© The Author(s), under exclusive license to Springer Nature Switzerland AG 2025
R. Allain, *Modeling Waves with Numerical Calculations Using Python*, Synthesis
Lectures on Wave Phenomena in the Physical Sciences,
https://doi.org/10.1007/978-3-031-78291-6_5

With f being the frequency. For a wave with a velocity v, this must be equal to the product of the wavelength and frequency which we can also write in terms of the wave number and angular frequency.

$$v = \lambda f = \frac{\omega}{k}$$

This particular function of y would be equivalent to a sine wave traveling in the positive x-direction. We could also have a function traveling in the negative x-direction.

$$y_2(x, t) = A \sin(kx + \omega t)$$

Is it possible to have both a wave traveling in the positive x-direction as well as a wave traveling in the negative x-direction? Suppose that we have a superposition of these two solutions.

$$y_1(x, t) = A \sin(kx - \omega t)$$

$$y_2(x, t) = A \sin(kx + \omega t)$$

$$y(x, t) = y_1(x, t) + y_2(x, t)$$

Let's assume that both y_1 and y_2 are solutions to the wave equation. Plugging in y(x, t) gives the second time derivative as

$$\frac{\partial^2}{\partial t^2}(y_1 + y_2) = \frac{\partial^2 y_1}{\partial t^2} + \frac{\partial^2 y_2}{\partial t^2}$$

For the space derivative:

$$v^2 \frac{\partial^2}{\partial x^2}(y_1 + y_2) = v^2 \left(\frac{\partial^2 y_1}{\partial x^2} + \frac{\partial^2 y_2}{\partial x^2} \right)$$

And we can see that:

$$\frac{\partial^2 y}{\partial t^2} = v^2 \frac{\partial^2 y}{\partial x^2}$$

So that this superposition is still a solution. Suppose we have two waves on a string with the same wave number (k) and angular velocity (ω) and amplitude (A) but traveling in opposite directions. We can get an idea about the behavior of this string by looking at a plot of the sum of these two waves for different time intervals (or alternatively you could make an animated graph). Here's what that looks like.

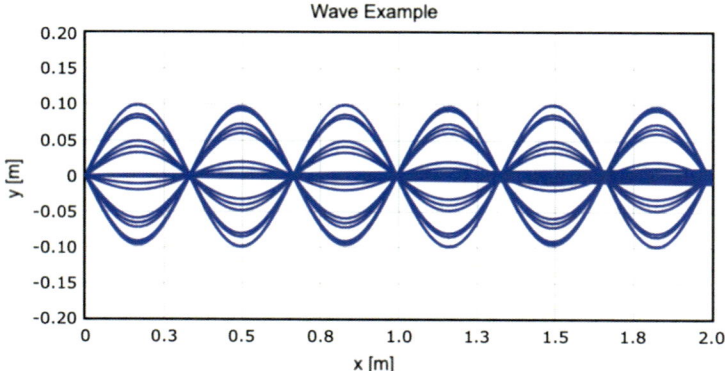

Each one of these curves represent the shape of the wave at different times. You can notice that it does oscillate. However, there are some locations at which the displacement of the string is zero for all values of time. These are stationary nodes. The distance from one node to the next is $\frac{\lambda}{2}$ or half a wavelength. This is indeed a standing wave.

We can get a mathematical model for a standing wave with a little bit of a trig identity. If the two traveling waves have the same amplitude, then we can write this as:

$$y(x, t) = y_1(x, t) + y_2(x, t) = A[\sin(kx - \omega t) + \sin(kx + \omega t)]$$

Since we now have the sum of two sine functions, we can use the following identity:

$$\sin \alpha + \sin \beta = 2 \sin \left(\frac{\alpha}{2} + \frac{\beta}{2} \right) \cos \left(\frac{\alpha}{2} - \frac{\beta}{2} \right)$$

where α and β are two functions (in this case $\alpha = kx - \omega t$ and $\beta = kx + \omega t$). We can now write the sum of the waves as:

$$y(x, t) = 2A \sin \left(\frac{kx - \omega t}{2} + \frac{kx + \omega t}{2} \right) \cos \left(\frac{kx - \omega t}{2} - \frac{kx + \omega t}{2} \right)$$

Which leads us to:

$$y(x, t) = 2A \sin(kx) \cos(-\omega t)$$

This gives us a product of functions. The first part of the function ($\sin(kx)$) only depends on space (x) and the second part ($\cos -(\omega t)$) only depends on time. We can think of this as space wave with an amplitude that changes with time.

Now consider the case with a string of length L and a wave number $k = \frac{\pi}{L} = \frac{2\pi}{\lambda}$. Let's look at the y value at the two ends of the string. At $x = 0$, we have:

$$y(0, t) = 2A \sin(0) \cos(-\omega t) = 0$$

And at x = L:

$$y(L, t) = 2A \sin(kL) \cos(-\omega t) = 0$$

For any value of t, the displacement at these two points will be zero since $\sin(\pi) = 0$. These will be nodes in which the wave doesn't oscillate. Of course that fits perfectly for our string with the two ends fixed and not able to move.

However, this won't work for any wave. This particular function requires that the wavelength is twice the length of the string (we call this the fundamental frequency). If the wavelength is equal to the length of the string, our solution still works. In this case there will be three zero points (nodes). They will be at x = 0, L/2, and L.

Now let's model replicate this standing wave with our ball and spring model of a wave. If we start with our previous model, we only need to change the initial positions of the individual balls. Looking at our standing wave equation, we get the following function at t = 0.

$$y(x, 0) = 2A \sin(kx) \cos(0) = 2A \sin(kx)$$

We can set the initial positions of the balls in the form of a sine wave with a wavelength of 2L.

```
30  for b in balls:
31      b.pos.y = .05*sin(2*pi*(b.pos.x+L/2)/L)
32
```

Notice that the left end of the balls list started at x = − L/2, so we need to adjust for that. Also, at this time each individual ball is at its maximum position. The instantaneous velocity at this time is zero thus each ball will have an initial momentum of < 0, 0, 0 > kg * m/s. Here's a snapshot of the output.

For the finite difference method, we just need to make a similar change to the code by setting the initial position of the elements.

```
38  for i in range(len(x)):
39      yold[i] = 0.05*sin(2*pi*x[i]/L)
40      ynow[i] = 0.05*sin(2*pi*x[i]/L)
41
```

Again, the initial velocity is zero for each element. The output gives us the following.

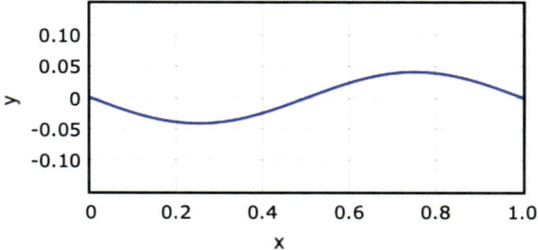

Yes, it's a standing wave.

The One Dimensional Wave Function

Now that we have some numerical methods to deal with a 1 dimensional wave on a string, let's use those same ideas for quantum mechanics. We can describe the behavior of quantum particles (like electrons) using the wave function and Schrödinger's equation.

This is not a lesson on one dimensional quantum mechanics but rather methods to solve this one particular equation.

$$i\hbar \frac{\partial \Psi}{\partial t} = -\frac{\hbar^2}{2m} \frac{\partial^2 \Psi}{\partial x^2} + V$$

Here, $\Psi(x, t)$ is the one dimensional wave function that can be used to describe the probability density for a particle. In this equation, $\hbar$ is a constant and m is the mass of the particle in a region with a potential function ($V(x)$)–we are going to assume that the potential does not change with time.

It looks very similar to our displacement as a function of time ($y(x, t)$) for a wave on a string. However, the Schrödinger equation is not a wave equation. Just look at the wave equation.

$$\frac{\partial^2 y}{\partial t^2} = v^2 \frac{\partial^2 y}{\partial x^2}$$

Even if the potential was zero, there are two big difference between the Schrödinger equation and the wave equation. First, the Schrödinger equation depends on the first derivative with respect to time. The second difference is that the Schrödinger equation includes an imaginary number (i) such that our solutions can be complex numbers. From now on, I'm going to reference the Schrödinger equation as SE (because it's way easier to type).

© The Author(s), under exclusive license to Springer Nature Switzerland AG 2025
R. Allain, *Modeling Waves with Numerical Calculations Using Python*, Synthesis
Lectures on Wave Phenomena in the Physical Sciences,
https://doi.org/10.1007/978-3-031-78291-6_6

6.1   Infinite Square Well

Let's start with a simple case and then make it more complicated. The easiest problem in quantum mechanics is a particle in a one dimensional infinite square well. The idea is that there is a particle of mass m confined to a region on the x-axis between x = 0 and x = a. Outside of that region V is equal to infinity so that it's impossible for the particle exist in there.

If we limit our solution to the region x = 0 to x = a, then the SE becomes:

$$i\hbar\frac{\partial\Psi}{\partial t} = -\frac{\hbar^2}{2m}\frac{\partial^2\Psi}{\partial x^2} + V(x)\Psi(x, t)$$

We can make the assume that the wave function is a product of a function that only depends on space and a function that only depends on time.

$$\Psi(x, t) = \psi(x)f(t)$$

Putting this into the SE gives:

$$i\hbar\frac{\partial\psi(x)f(t)}{\partial t} = -\frac{\hbar^2}{2m}\frac{\partial^2\psi(x)f(t)}{\partial x^2} + V(x)\psi(x)f(t)$$

Since these are partial derivatives (for both space and time) we can pull the time function out of the space derivative and pull the space function out of the time derivative.

$$i\hbar\psi(x)\frac{\partial f(t)}{\partial t} = -\frac{\hbar^2}{2m}f(t)\frac{\partial^2\psi(x)}{\partial x^2} + V(x)\psi(x)f(t)$$

Now we can divide both sides of the equation by $f(t)\psi(x)$.

$$i\hbar\frac{1}{f(t)}\frac{\partial f(t)}{\partial t} = -\frac{\hbar^2}{2m}\frac{1}{\psi(x)}\frac{\partial^2\psi(x)}{\partial x^2} + V(x)$$

This gives an equation with the left side only depending on time and the right side only depending on position (x). The only way for this equation to work is if both sides are equal to a constant. Let's call this constant E (yes, for energy).

Looking at the time part, we have the following differential equation.

$$\frac{1}{f(t)}i\hbar\frac{\partial f(t)}{\partial t} = E$$

$$\frac{\partial f(t)}{\partial t} = -\frac{iE}{\hbar}f(t)$$

This isn't too difficult to solve. We can guess a function for f such that when we take the partial derivative of f with respect to time, we get the same function with a constant.

Here's a solution.

$$f(t) = e^{-iEt/\hbar}$$

You can (and should) check that this function satisfies the differential equation. This time function will be very important later. But let's jump over to the space part of the equation.

$$-\frac{\hbar^2}{2m}\frac{1}{\psi(x)}\frac{\partial^2\psi(x)}{\partial x^2} + V(x) = E$$

$$-\frac{\hbar^2}{2m}\frac{\partial^2\psi(x)}{\partial x^2} + V(x)\psi(x) = E\psi(x)$$

Going back to the infinite square well, we know that $V(x) = 0$ inside the well. Let's also introduce a new variable (just to make things cleaner) for all the constants.

$$k = \frac{\sqrt{2mE}}{\hbar}$$

So that inside the well, we have the following differential equation.

$$\frac{\partial^2\psi(x)}{\partial x^2} = -k^2\psi(x)$$

That's not too difficult to solve. We can again make a guess for the function $\psi(x)$ and see if it works. It needs to be a function such that when we take the partial derivative twice we get the same function with a negative constant. The sine and cosine function both will work. Let's try the following:

$$\psi(x) = A\sin(kx) + B\cos(kx)$$

With constants A and B. Now we can apply our boundary conditions. Since the particle in the one dimensional box can only exist between $x = 0$ and $x = a$, the probability and thus the wave function must be zero at these points.

$$\psi(0) = 0$$

$$\psi(a) = 0$$

Plugging into our solution for $x = 0$, we get:

$$\psi(0) = A\sin(0) + B\cos(0) = 0$$

Since $\sin(0) = 0$ and $\cos(0) = 1$ this means that B must be equal to zero. For the other boundary condition:

$$\psi(a) = A\sin(ka) = 0$$

There are two ways to make this condition valid. First, it's possible that $A = 0$. However, with that method our function is just zero. That's not good. The second method is to let $\sin(ka) = 0$. This will be true for the following:

$$\psi(a) = A\sin(ka) = 0$$

$$ka = n\pi$$

$$k = \frac{n\pi}{a}$$

where n is an integer value so that k can only have discrete values. But recall that k is related to the energy (E). This mean that E can also only have discrete values.

$$k = \frac{n\pi}{a} = \frac{\sqrt{2mE}}{\hbar}$$

$$E_n = \frac{n^2\pi^2\hbar}{2ma^2}$$

With each quantized value of energy, we have a solution for $\psi(x)$. This can be expressed as the following function.

$$\psi(x) = A\sin(k_n x)$$

where I have denoted the variable k with the subscript n to emphasize that it depends on the integer value of n. The value of A can be determined through normalization. This says that the total probability of find the particle somewhere in the infinite well is equal to 1. However, since we want to really focus on numerical methods to find the wave equation, we can just leave this as A.

6.2   Shooting Method

Now suppose that you want to find a numerical solution for the space part of the wave function for a particle in an infinite well. Think back to the Euler method for a mass on a spring from the first python calculation. In general, we had a differential equation of the form:

$$\frac{d^2x}{dt^2} = -\frac{k}{m}x$$

We solved this numerically by using small time intervals (Δt) and assuming the acceleration was constant. With that we could update the velocity and use that to update the position.

$$v_2 = v_1 + a\Delta t$$

$$x_2 = x_1 + v\Delta t$$

Could we do something similar for the space part of the SE (also called the time-independent Schrodinger equation–the TISE)? Well, we can certainly solve for the second derivative of the wave function. OK, let's make a quick notation comment. Since we are dealing with partial derivatives with respect to position, I'm going to use the following convention.

$$\frac{\partial \psi}{\partial x} = \psi'$$

$$\frac{\partial^2 \psi}{\partial x^2} = \psi''$$

With that, the TISE looks like this:

$$\psi'' = -\psi \frac{2m}{\hbar^2}(E - V)$$

Isn't it possible to use the same numerical method for a mass on a spring with the wave function? Maybe. If we break the space into small space pieces (Δx) then we can write the second derivative as a finite difference.

$$\psi'' = \frac{\Delta \psi'}{\Delta x}$$

With this, we could solve for the value of Ψ' at the end of the space step.

$$\psi'_2 = \psi'_1 + \psi''\Delta x$$

That looks just like the solution for a mass on a spring. But there's one big problem. We actually don't know the initial value of Ψ' like we knew the initial velocity for a mass and spring. Instead, of initial conditions, we have boundary conditions. We know the value of $\psi(0)$ and $\psi(a)$.

This is where the shooting method comes into play. The best way to explain the shooting method is with a basketball analogy. Imagine that you have a ball and you want to shoot it so that it lands in the basket. However, you aren't sure how fast you need to throw the ball in order to make the shot. So, you just start with some initial velocity and see where the ball lands. If the ball lands short, you get the ball and throw it again–but this time with a slightly higher launch speed. You keep doing this until the ball hits the basket and then you know the correct launch speed.

For the one dimensional infinite square well, we will do something very similar. There are actually two things that we don't know to do an Euler method calculation. We don't know the initial slope (Ψ') and we don't know the value of E (the energy). Let's just pick $\Psi' = 1$ and $E = 0$. With that we can just do a normal numerical calculation. Here is the code for that calculation.

```
 3  g1 = graph(title="1D Square Well", xtitle="x",
 4  ytitle="psi", width=500, height=250)
 5  f1 = gcurve()
 6
 7  x = 0
 8  dx = 0.01
 9  psi = 0
10  dpsi = 1
11  E = 0
12  m = 1
13  hbar = 1
14  a = 1
15
16  while x<a:
17      ddpsi = -2*m*E*psi/hbar**2
18      dpsi = dpsi + ddpsi*dx
19      psi = psi + dpsi*dx
20      x = x + dx
21      f1.plot(x,psi)
22
23
```

Notice the weird units. In order to keep the values under control I'm going to use 1 for the mass of the particle and 1 for the value of $\hbar$. When we run this, we get the following output.

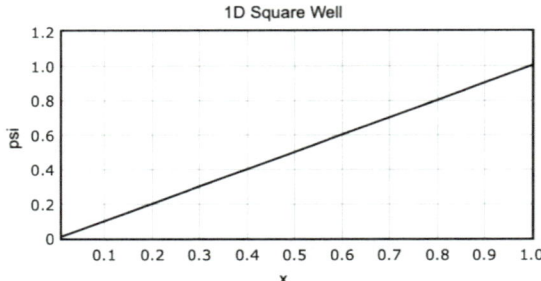

That clearly doesn't meet the boundary condition. If you increase the value of E to 1, the output looks like this:

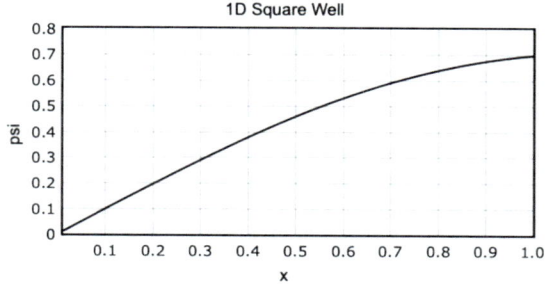

That's getting closer, so we just need to keep increasing the value of E until the boundary condition at x = 1 is satisfied. It's possible to manually change the value of E until you get a solution. The other option is to make it automatic. Here is the code for that.

```
 3 g1 = graph(title="1D Square Well", xtitle="x",
 4 ytitle="psi", width=500, height=250)
 5 f1 = gcurve()
 6
 7 x = 0
 8 dx = 0.01
 9 psi = 0
10 dpsi = 1
11 E = 0
12 dE = 0.01
13 m = 1
14 hbar = 1
15 a = 1
16
17 psifinal=1
18 while psifinal>0:
19     rate(500)
20     psi = 0
21     dpsi = 1
22     x = 0
23     f1data = []
24     while x<a:
25         ddpsi = -2*m*E*psi/hbar**2
26         dpsi = dpsi + ddpsi*dx
27         psi = psi + dpsi*dx
28         x = x + dx
29         f1data = f1data +[[x,psi]]
30     f1.data = f1data
31     psifinal = psi
32     E = E + dE
33 print("E = ",E)
34
```

Let's go over the important parts of this nested loop code. First, in line 12 I need to define my energy step size (dE = 0.01). If the step size is smaller you will get a better value for the final energy however, it will take longer to find that solution.

In line 17 I have a new variable called "psifinal" and it's initially set to a value of 1. Notice that there are two loops. The outer loop runs as long as psipfinal is positive. As we increase the value of the energy, the final value for ψ decreases and gets closer to zero at the boundary. If we keep on increasing the value of E, the final value of ψ will eventually become negative. So the variable psifinal is a way to keep track of this final condition.

In the beginning of the outer loop we need to do a few things. The values for psi and dpsi need to be reset (since they are changed in the Euler calculation part of the shooting method). The value of x also needs to be reset to zero. Line 23 is used to create an animated graph showing psi as a function of position. In this line, we reset the list fldata to be an empty list so that we can add new values.

The inner loop (staring at line 24) is just the same code as in the example when the value for energy was changed manually. When the outer loop ends (with the boundary condition satisfied) the value for energy is also printed. Here is the final output after the code has run.

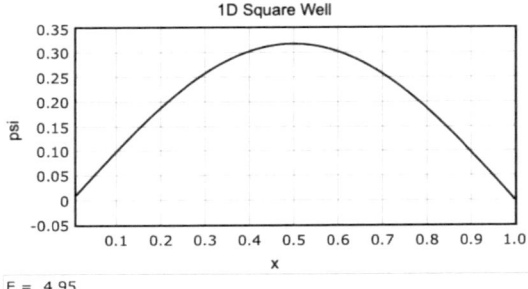

E = 4.95

This is indeed a sine function. If you calculate the energy for n = 1 in the analytical solution, it's very close to 4.95. So, it seems this works. Recall that we started with $\Psi'(0) = 1$—an arbitrary choice. It's possible to get an exact value for Ψ' by assuming the wave function is normalized over this region.

6.3   Infinite Square Well with the Finite Difference Method

It's possible to find the wave function inside an infinite square well using the finite differ-ence method–in a way that's mostly the same as the method used for a wave on a string with fixed ends.

Just like a wave on a string, let's start by breaking the wavefunction into finite space elements Δx. This will give a finite number of values for the wavefunction which we can label as ψ_1, ψ_2, ψ_3…and so on. Also, like before we can define the first partial derivative. Suppose we look at the wave function elements ψ_{i-1} to ψ_{i+1} that covers a space of $2\Delta x$. The slope (and thus the first partial derivative) would be calculated as:

$$\frac{\partial \psi_i}{\partial x} = \psi_i' = \frac{\psi_{i+1} - \psi_{i-1}}{2\Delta x}$$

Note that I am using the "prime" notation to indicate a partial derivative with respect to space. Using this same idea, the second partial derivative with respect to x would be a change in the first partial derivative. If we want the second derivative of ψ_i we can use the change in the slope at the "half intervals" of $\psi'_{i+\frac{1}{2}}$ and $\psi'_{i-\frac{1}{2}}$ $\psi'_{i-\frac{1}{2}}$

$$\psi''_{i_i} = \frac{\psi'_{i+1/2} - \psi'_{i-1/2}}{\Delta x}$$

Using the definition for the first derivative:

$$\psi''_i = \frac{\frac{\psi_{i+1} - \psi_i}{\Delta x} - \frac{\psi_i - \psi_{i-1}}{\Delta x}}{\Delta x}$$

$$\psi''_i = \frac{\psi_{i+1} - \psi_i - \psi_i + \psi_{i-1}}{\Delta x^2}$$

$$\psi''_i = \frac{\psi_{i-1} - 2\psi_i + \psi_{i+1}}{\Delta x^2}$$

This is quite useful. We can calculate the second derivative for the i-th element of the wave function by just knowing the values before and after that element. Now with this new definition of the second partial derivative, we can put this into the TISE.

$$\psi''_i = -\frac{2mE}{\hbar^2}\psi_i$$

$$\psi''_i = \frac{\psi_{i-1} - 2\psi_i + \psi_{i+1}}{\Delta x^2} = -\frac{2mE}{\hbar^2}\psi_i$$

We can solve this equation for ψ_i:

$$2\psi_i - \frac{2mE}{\hbar^2}\Delta x^2 \psi_i = \psi_{i+1} + \psi_{i-1}$$

$$\psi_i = \frac{\psi_{i+1} + \psi_{i-1}}{2 - \frac{2mE\Delta x^2}{\hbar^2}}$$

That equation looks complicated–but it's really not too bad. It calculates the value of ψ_i just using the values of ψ before and after that particular value. But how can you find the value at the location "i" if you don't know the values at $i - 1$ and $i + 1$? Yes, that's a problem. Here's a recipe for how we can use this method to determine the wave function in an infinite square well.

- Break the space into finite elements (Δx) in size.
- Make a "rough guess" for the values of ψ for all elements at the finite spaces.

- Now that I have values for ψ (even if they are wrong) I can use those values to find BETTER values of ψ. I can just keep repeating this process until I get better values for the wave function.

Of course there is a small problem (that's actually not a problem). Since we are updating the values of ψ using values to the left and right of this point, how do we calculate the first value at ψ_0? How do we calculate the value for the last element? We can't use this same method because there aren't values of ψ before the first value. But don't worry–we don't need to calculate these values of ψ since they are the boundary conditions for our infinite square well.

Now we are ready to code this method in python. First, we need to set up our lists for the values of x and the wave function. Here's that part of the code (followed by comments).

```
13
14 xp=[]
15 psi=[]
16 psi2=[]
17 xp = xp + [x]
18 psi = psi +[0]
19 psi2 = psi2 +[0]
20 dx = 0.05
21
22 while x<=a:
23     x = x + dx
24     xp = xp + [x]
25     psi = psi +[.4]
26     psi2 = psi2 +[0.4]
27 psi[-1] =0
28 psi2[-1]=0
29 fdata=[]
30 for i in range(len(xp)):
31     fdata=fdata + [[xp[i],psi[i]]]
32 f1.data=fdata
33
```

- The first step is to make some lists (in lines 14–16). xp is a list of the values of x, then I have psi for the values of ψ. But I need ANOTHER list of values for ψ called psi2. These will be the updated values based on the previous values.
- In order to start filling in the values for these three lists, I set their first values in lines 17–19. I'm using x as an actual variable with xp the list of x-values such that the initial value of x is zero. For the two wave functions, the first value must be zero in order to satisfy the boundary conditions.
- In line 22–26, the code goes through and fills in values into the three lists. Notice that both of the lists for the wave function (psi and psi2) start off with a constant value of 0.4 except for the first element (which was set to zero).

- In line 27–28, I set the values of psi and psi2 at the other boundary condition (which would be the last element).
- Finally, lines 30–32 are just to plot this initial wave function.

Here's what that starting wave function looks like.

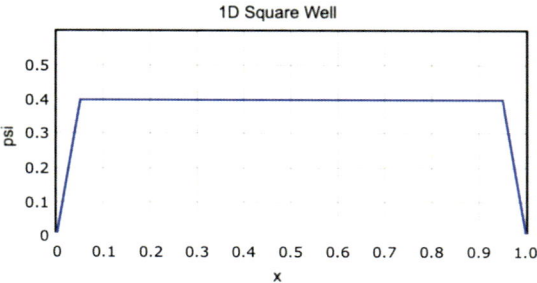

Now for the second part of the python code.

```
38  n = 0
39  N = 50
40  while n<N:
41      rate(50)
42      fdata=[]
43      for i in range(1,len(xp)-1):
44          psi2[i]=(psi[i+1]+psi[i-1])/(2-2*m*E*dx**2/hbar**2)
45      for i in range(len(psi)):
46          psi[i]=psi2[i]
47
48      for i in range(len(psi)):
49          fdata=fdata+[[xp[i],psi[i]]]
50
51      f1.data=fdata
52      n=n+1
53
```

- Lines 38–39. These are "counters" for how many times the code will go over the wave function values and calculate new ones. In this cases, I'm going to run $N = 50$ iterations with $n = 0$ being the counter.
- Line 40–52 is a loop over the number of iterations. In my version of the code, I make an animated graph showing how the wave function changes with each pass. The rate (50) just tells the program to only do 50 loops every second. The animated graph is created in lines 48–49 and then plotted in line 51. Remember to reset the plotting list in line 42.
- Lines 43–44 is the main calculation. This for loop goes over the elements in the list starting from the second item (index of 1) up to one less than the length of the list. That means that we won't change the first or last item in the list of wave functions. Line 44 is the same thing as the TISE above solving for the value of ψ_i. Notice that we assign these values to the NEW list, psi2 using the old values of psi.

- Lines 45–46 take the old values of psi and set them to the new values of psi. This is needed for the next pass through this loop.

After just 50 iterations, we get the following output for the wave function.

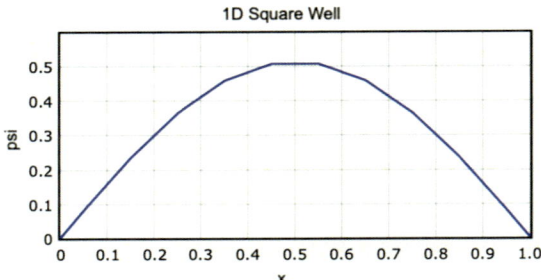

This looks very similar to our other solutions for the wave function in a one dimensional infinite square well.

However, this method isn't perfect. It doesn't directly give us the values of the quantized energy (E). Also, if you want the next solution (for $n = 1$) you would need to start with a different "guess" wave function. If know that it will look something like a full sine wave, you could start with a guess that looks like this:

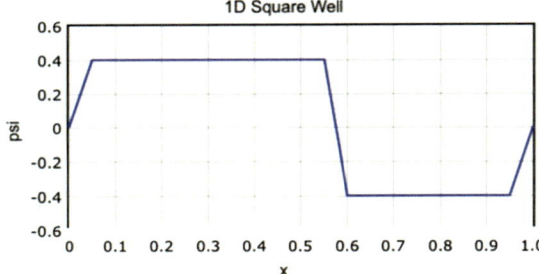

Also setting the value of E to something close to what we expect, the final output looks like this:

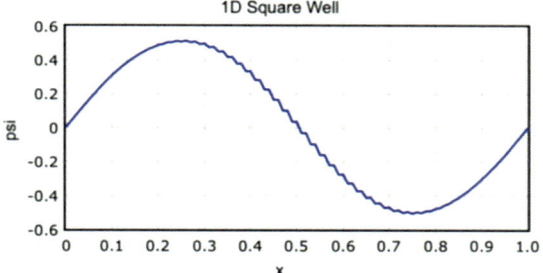

The Eigenvalue Problem

There's a much better method using the finite difference method for a 1D square well. We can still break the wave function into finite elements (ψ_i) and we can still use the same definition for the second partial derivative.

Just for simplicity, imagine that the wave function is split into 5 elements (ψ_0 through ψ_4) with $\psi_0 = 0$ and $\psi_4 = 0$ being set as boundary conditions. That means that I can write the TISE for the three elements in the middle. Let's make one small modification. In the previous versions of the TISE, we assumed that the potential (V) was zero, It will be more useful later if we leave V in the equation.

$$-\frac{\hbar^2}{2m}\frac{\partial^2 \psi}{\partial x^2} + V\psi = E$$

Here the potential (V) can also have finite values at different x-positions. With the finite difference version of the second derivative, the i-th element of the wave function would be:

$$-\frac{\hbar^2}{2m\Delta x^2}(\psi_{i+1} - 2\psi_i + \psi_{i-1}) + V_i\psi_i = E\psi_i$$

$$-\frac{\hbar^2}{2m\Delta x^2}(\psi_{i+1} + \psi_{i-1}) + \left(\frac{\hbar^2}{m\Delta x^2} + V_i\right)\psi_i = E\psi_i$$

If we use this for our three points in the middle of the square well, we get three of these differential equations.

$$-\frac{\hbar^2}{2m\Delta x^2}\psi_0 + \left(\frac{\hbar^2}{m\Delta x^2} + V_1\right)\psi_1 - \frac{\hbar^2}{2m\Delta x^2}\psi_2 = E\psi_1$$

© The Author(s), under exclusive license to Springer Nature Switzerland AG 2025
R. Allain, *Modeling Waves with Numerical Calculations Using Python*, Synthesis Lectures on Wave Phenomena in the Physical Sciences,
https://doi.org/10.1007/978-3-031-78291-6_7

$$-\frac{\hbar^2}{2m\Delta x^2}\psi_1 + \left(\frac{\hbar^2}{m\Delta x^2} + V_2\right)\psi_2 - \frac{\hbar^2}{2m\Delta x^2}\psi_3 = E\psi_2$$

$$-\frac{\hbar^2}{2m\Delta x^2}\psi_2 + \left(\frac{\hbar^2}{m\Delta x^2} + V_3\right)\psi_3 - \frac{\hbar^2}{2m\Delta x^2}\psi_4 = E\psi_3$$

These are three linear equations for the variables ψ_1, ψ_2 and ψ_3. Since these are linear equations we can write this as the following matrix operation.

$$\frac{\hbar^2}{2m\Delta x^2}\begin{pmatrix} -2-kV_1 & 1 & 0 \\ 1 & -2-kV_2 & 1 \\ 0 & 1 & -2-kV_3 \end{pmatrix}\begin{pmatrix} \psi_1 \\ \psi_2 \\ \psi_3 \end{pmatrix} = E\begin{pmatrix} \psi_1 \\ \psi_2 \\ \psi_3 \end{pmatrix}$$

Here I have introduced the constant k (for simplicity) where:

$$k = \frac{2m\Delta x^2}{\hbar^2}$$

You should check that this works. However, it allows us to write this system of equations in a very organized format. It actually turns the TISE into this equation:

$$kM\vec{\psi} = E\vec{\psi}$$

where $\vec{\psi}$ is the vector consisting of the finite elements of the wave function (in this example, that's just three items) and M is the 3×3 matrix consisting of a combination of the second derivative and the potential. When this matrix operates on the wave function vector, we get the same vector back with a scalar constant and that constant is the energy (E). In fact, the product of kM is the Hamiltonian. This gives us the following representation of the TISE.

$$H\vec{\psi} = E\vec{\psi}$$

This is exactly the eigenvalue problem. It not only appears here in a quantum system but also in the principle axis of rotation for a rotating rigid object and in the coupled oscillations for a multi-mass system.

Notice that if you increase the number of elements in the wave function to 50, the matrix M would be 50×50. However, many of those elements in the 50×50 matrix would in fact be zeros. The only non-zero values will be along the diagonal of the matrix and the two off-diagonals. So this is actually a tridiagonal matrix.

7.1   A 2 × 2 Eigenvalue Example

But how do you solve the eigenvalue problem? Let's just look at a generic eigenvalue problem that looks like this:

$$A\vec{r} = \lambda\vec{r}$$

where A is some matrix operator and λ is a scalar eigenvalue. We can operate the right side of the equation with the identity matrix (1) and not change anything. But that will make the right side an nxn matrix just like the left side that we can subtract it from both sides of the equation.

$$A\vec{r} = 1\lambda\vec{r}$$

$$A\vec{r} - 1\lambda\vec{r} = \vec{0}$$

Here, $\vec{0}$ is a $1 \times n$ dimensional zero vector (with all the components of the vector equal to zero). Factoring out the vector r:

$$(A - 1\lambda)\vec{r} = \vec{0}$$

This will only have non-trivial solutions if the determinant of the matrix $(A - 1\lambda)$ is equal to zero.

$$det|A - 1\lambda| = 0$$

From the determinant of this matrix, we can find values for λ (the eigen values) and use those to find the vector $(\vec{r})$ that satisfy this equation. Before moving to larger matrices, let's do a simple example with a 2×2 matrix for the operator A. Let's use the following matrix.

$$A = \begin{pmatrix} 2 & 3 \\ 2 & 1 \end{pmatrix}$$

For the 2×2 case, the identity matrix is:

$$1 = \begin{pmatrix} 1 & 0 \\ 0 & 1 \end{pmatrix}$$

This means we want to take the determinant of the matrix

$$A - 1\lambda = \begin{pmatrix} 2 - \lambda & 3 \\ 2 & 1 - \lambda \end{pmatrix}$$

The determinant of 2×2 is fairly straight forward, it's just the multiplication of the two diagonals subtracted. That gives the following equation when the determinant is set to zero.

$$det|A - 1\lambda| = (2 - \lambda)(1 - \lambda) - 6 = 0$$

If we expand the parenthesis, we get:

$$2 - 2\lambda - \lambda + \lambda^2 - 6 = 0$$

$$\lambda^2 - 3\lambda - 4 = 0$$

This is a second order polynomial. We could of course determine the values for λ using the quadratic equation. However, in this case it's also possible to solve by factoring.

$$\lambda^2 - 3\lambda - 4 = (\lambda - 4)(\lambda + 1) = 0$$

This gives the following two solutions:

$$\lambda_1 = 4$$

$$\lambda_2 = -1$$

But what about the eigenvectors? Let's start with the vector that goes with the eigenvalue of $\lambda_1 = 4$. Using that value, our original eigenvalue equation becomes:

$$A\vec{r} = 4\vec{r}$$

$$\begin{pmatrix} 2 & 3 \\ 2 & 1 \end{pmatrix} \begin{pmatrix} x \\ y \end{pmatrix} = 4 \begin{pmatrix} x \\ y \end{pmatrix}$$

If we operate the matrix on the x, y vector we get the following two equations.

$$2x + 3y = 4x$$

$$2x + y = 4y$$

These are not independent equations. If we just look at the first equation, we can simplify to:

$$3y = 2x$$

We can now pick values for x and y that satisfy this equation. Let's use $y = 2$ and $x = 3$. That means that the following vector (we can call it $\vec{r_1}$) should satisfy the eigenvalue equation.

$$\begin{pmatrix} x \\ y \end{pmatrix} = \begin{pmatrix} 3 \\ 2 \end{pmatrix}$$

Let's check if this actually works

$$\begin{pmatrix} 2 & 3 \\ 2 & 1 \end{pmatrix}\begin{pmatrix} 3 \\ 1 \end{pmatrix} = \begin{pmatrix} 12 \\ 8 \end{pmatrix} = 4\begin{pmatrix} 3 \\ 2 \end{pmatrix}$$

Yes, the vector (3, 2) is indeed an eigenvector solution to the original equation. Of course ANY vector in the same direction would also work. One thing we often do is to find a normalized eigenvector. The magnitude of this vector is:

$$|\vec{r_1}| = \sqrt{3^2 + 2^2} = \sqrt{13}$$

This gives a normalized eigenvector of:

$$\vec{r_1} = \frac{1}{\sqrt{13}}\begin{pmatrix} 3 \\ 2 \end{pmatrix} = \begin{pmatrix} 0.832 \\ 0.555 \end{pmatrix}$$

Repeating this process for the other eigenvalue gives the following solution.

$$\vec{r_2} = \begin{pmatrix} -1 \\ 1 \end{pmatrix}$$

If we normalize it, we get:

$$\vec{r_2} = \begin{pmatrix} -0.707 \\ 0.707 \end{pmatrix}$$

Those are the two solutions to the original equation. Of course this was a 2 × 2 matrix operator. Imagine a 10 × 10 matrix. The steps would be the same but there would be many more steps.

With some many simple, but relatively straight forward steps in a calculation like this, it's not so great for humans. However, it's perfect for a computer to solve these problems. That's exactly what we will do–use python to solve the eigenvalue problem.

7.2   Jupyter Notebooks and Numpy

For the previous calculations in python, we we able to use Web VPython. However, this implementation of python does not have native method to deal with matrices and matrix operations. For that, we will need to run "real" python.

There are many options to run python. One of the common methods is to run a Jupyter notebook. You can install this on your own system (Anaconda is a good option—https://www.anaconda.com/) or you can run it online using Google Colab (https://colab.research.google.com/). The colab is a nice option since you can run it without installing anything. This is perfect if you have a tablet instead of a computer.

The key to Jupyter notebooks is that it runs in a web browser and let's you create different cells. These cells can be either a chunk of python code or it could be some type of text. This allows you to write and run a particular code in a cell and then write a description in the following cell. I'm not going to give a full tutorial on how to use notebooks but instead focus on the code.

The other important aspect of python (real python) is its dependence on modules. We will need to import different modules that have the functions that we want to use. In this case we need just one module (for now)–numpy. This is a module with just about every numerical function that you will need.

Let's just get to it. I want to use numpy to find the normalized eigenvectors for the same 2×2 operator above. Since modules like python aren't already there, we need to load them.

```
In [1]:    1  import numpy as np
```

Here's I'm giving the name of numpy as np. This is the most common name for numpy, but you could give it any name if that makes you happy. If you want to use a function from the numpy module, you can reference it as np.

As an example, suppose I have the variable x = 3 and I want to take the square root of that variable. The square root function isn't in plain python, so we will have to use it in numpy. Here's what that looks like.

```
In [2]:    1  x = 3
           2  y = np.sqrt(x)
           3  print(y)
1.7320508075688772
```

If you want to see what mathematical functions are in numpy, you can check out the documentation—https://numpy.org/doc/stable/reference/routines.math.html.

Now, let's use numpy to solve the eigenvalue problem. The first step is to define our A operator as something like a matrix. What we can do is to make an array of 2 arrays. In numpy, an array is much like a list but it's more powerful.

```
In [3]:   1  A = np.array([[2,3],[2,1]])
          2  print("A = ",A)

          A =  [[2 3]
               [2 1]]
```

There is a function in numpy that calculates both the eigenvalues and eigenvectors for this matrix.

```
In [4]:   1  a,r = np.linalg.eig(A)
          2  print("a = ",a)
          3  print("r = ",r)

          a =  [ 4. -1.]
          r =  [[ 0.83205029 -0.70710678]
               [ 0.5547002   0.70710678]]
```

The numpy function is called linalg.eig and it returns two values. It produces the eigenvalues (which will set to the variable "a") and the eigenvectors to the variable "r".

The two eigenvalues are exactly the same as the ones we found analytically. However, the eigenvectors might look different. This function returns the eigenvectors as a matrix with the vectors as column vectors. However, if we want to print just one of the vectors we need to take the transpose of this matrix. Here is the first eigenvector.

```
In [6]:   1  print("r1 = ",r.T[0])

          r1 =  [0.83205029 0.5547002 ]
```

Notice that we take the transpose (T) and then reference the first element ([0]). This is indeed the same normalized vector that we found before. Here is the second normalized eigenvector.

```
In [7]:   1  print("r2 = ",r.T[1])

          r2 =  [-0.70710678  0.70710678]
```

Now that we have reproduced the eigen values and eigenvectors for a simple 2×2 matrix, we can use the same method for something more complicated–like our 1D infinite square well.

7.3   Eigenvalues for the 1D Infinite Square Well

In a previous section, I showed how to write the TISE using a finite difference derivative for an infinite square well with wave functions defined at just 5 different points (2 of those wave function values were boundary conditions).

But what if we want a better numerical solution? What about a solution by breaking the space into 10 values? What about 100 values? The same method used for a 3×3 matrix will work for a 10×10 or even 100×100–but it's going to be too messy to write it out. Let's just do this in python. Before getting to the calculations, we are going to need to import a second module. There's a useful module for creating graphs–it's called matplotlib.pyplot. Here's the starting code.

```
In [1]:    1  import numpy as np
           2  import matplotlib.pyplot as plt
```

Recall that we imported numpy as np (since that's the common name). The same is true for matplotlib.pyplot–pretty much every imports this as "plt". You are of course welcome to call it something else, but I wouldn't recommend it.

Next, let's enter our constants.

```
In [ ]:    1  a = 1
           2  m = 1
           3  hbar = 1
```

We are again going to use simple values for the mass (m), the width of the well (a), and $\hbar$. Now we need to create our array of x-values. For this, we can use the numpy.linspace function. This builds an array of values. Let me show you an example and then give a more detailed explanation.

```
In [3]:    1  N = 10
           2  x = np.linspace(0,a,N+1)
           3  print("x = ",x)

x =  [0.  0.1 0.2 0.3 0.4 0.5 0.6 0.7 0.8 0.9 1. ]
```

The linspace function takes 3 parameters: the starting value of the array, the final value of the array and the number of items in the array. In this case we want 10 "intervals" instead of 10 values. Starting at x = 0 and going to x = a for 10 intervals would mean 10 + 1 values. I know that can seem confusing–the best option is to just use the linspace function and try changing the different values to see what you get.

While we are looking at the x array, now would be an appropriate time to create the variable for Δx (which I will call dx in the code). It might be tempting to just set this equal to 0.1 or even a/N. However, I like to just explicitly define this dx value as the difference between the second and first x element.

```
In [4]:    1  dx = x[1]-x[0]
           2  print("dx = ",dx)

dx =  0.1
```

Now it's time to build our M matrix. Although it's possible code this manually as an array of lists like we did for the 2×2 case, there's an easier way. Here's the code for that followed by an explanation. Also, I am including print statements so that you can see what the values look like–but of course that's not required for the full calculation. We can build this matrix in different steps. Since it's a tridiagonal matrix, let's make a matrix with just a diagonal values of 1.

```
In [5]:   1  M = np.diag(np.ones(N-1))
          2  print(M)

          [[1. 0. 0. 0. 0. 0. 0. 0. 0.]
           [0. 1. 0. 0. 0. 0. 0. 0. 0.]
           [0. 0. 1. 0. 0. 0. 0. 0. 0.]
           [0. 0. 0. 1. 0. 0. 0. 0. 0.]
           [0. 0. 0. 0. 1. 0. 0. 0. 0.]
           [0. 0. 0. 0. 0. 1. 0. 0. 0.]
           [0. 0. 0. 0. 0. 0. 1. 0. 0.]
           [0. 0. 0. 0. 0. 0. 0. 1. 0.]
           [0. 0. 0. 0. 0. 0. 0. 0. 1.]]
```

There is a numpy function "diag" that creates a diagonal matrix. The values for the diagonals will be one. The function np.ones creates a bunch of values of "1"–in this case we are going to make $N - 1$ or 9 ones. Why not 10 values? Remember that we have 11 values of x (from 0 to a). You might think that we would have 11 values for our wave function also–but remember that the values for the wave function will be set at zero for both of the endpoints. That leaves just 9 values. This is in fact a 9×9 matrix. You can change this with the np.shape function.

```
In [6]:   1  print(np.shape(M))

          (9, 9)
```

Of course we don't actually want 1's on the diagonals, we want $- 2$'s. Here's a slightly different code for that.

```
In [7]:   1  M = np.diag(-2*np.ones(N-1))
          2  print(M)

          [[-2.  0.  0.  0.  0.  0.  0.  0.  0.]
           [ 0. -2.  0.  0.  0.  0.  0.  0.  0.]
           [ 0.  0. -2.  0.  0.  0.  0.  0.  0.]
           [ 0.  0.  0. -2.  0.  0.  0.  0.  0.]
           [ 0.  0.  0.  0. -2.  0.  0.  0.  0.]
           [ 0.  0.  0.  0.  0. -2.  0.  0.  0.]
           [ 0.  0.  0.  0.  0.  0. -2.  0.  0.]
           [ 0.  0.  0.  0.  0.  0.  0. -2.  0.]
           [ 0.  0.  0.  0.  0.  0.  0.  0. -2.]]
```

Next, we need to create the two off diagonal elements. Here's a test matrix with two off diagonals with values of 1.

```
In [9]:    1  A = np.diag(np.ones(N-2),1) + np.diag(np.ones(N-2),-1)
           2  print(A)

[[0. 1. 0. 0. 0. 0. 0. 0. 0.]
 [1. 0. 1. 0. 0. 0. 0. 0. 0.]
 [0. 1. 0. 1. 0. 0. 0. 0. 0.]
 [0. 0. 1. 0. 1. 0. 0. 0. 0.]
 [0. 0. 0. 1. 0. 1. 0. 0. 0.]
 [0. 0. 0. 0. 1. 0. 1. 0. 0.]
 [0. 0. 0. 0. 0. 1. 0. 1. 0.]
 [0. 0. 0. 0. 0. 0. 1. 0. 1.]
 [0. 0. 0. 0. 0. 0. 0. 1. 0.]]
```

The same np.diag function can make off diagonals. Using the value of 1 makes it above the diagonal and -1 is below. Notice that we have $N - 2$ values for these off diagonals. Now we can put it all together and make the Hamiltonian matrix.

```
In [11]:   1  M = np.diag(-2*np.ones(N-1))+np.diag(np.ones(N-2),1)+np.diag(np.ones
           2  k = -hbar**2/(2*m*dx**2)
           3  H = k*M
           4  print(H)

[[100. -50.  -0.  -0.  -0.  -0.  -0.  -0.  -0.]
 [-50. 100. -50.  -0.  -0.  -0.  -0.  -0.  -0.]
 [ -0. -50. 100. -50.  -0.  -0.  -0.  -0.  -0.]
 [ -0.  -0. -50. 100. -50.  -0.  -0.  -0.  -0.]
 [ -0.  -0.  -0. -50. 100. -50.  -0.  -0.  -0.]
 [ -0.  -0.  -0.  -0. -50. 100. -50.  -0.  -0.]
 [ -0.  -0.  -0.  -0.  -0. -50. 100. -50.  -0.]
 [ -0.  -0.  -0.  -0.  -0.  -0. -50. 100. -50.]
 [ -0.  -0.  -0.  -0.  -0.  -0.  -0. -50. 100.]]
```

With the Hamiltonian matrix, we just need to use the same linear algebra function to find the eigen values and vectors.

```
In [12]:   1  E, psi = np.linalg.eigh(H)
           2  print(E)

[   4.89434837  19.09830056  41.22147477  69.09830056 100.
  130.90169944 158.77852523 180.90169944 195.10565163]
```

Notice that I am using np.linalg.eigh instead of np.linalg.eig. Both functions will work, but in this case the eigh is more efficient (faster). I have printed out the energies. Since we have a 9×9 Hamiltonian matrix, we get 9 eigen values–these 9 values are the first 9 energy levels.

What about the eigenvectors? Instead of printing those, let's create a plot. The first issue that we should deal with are the number of values. Remember that our method assumes that the wave function is zero at $x = 0$ and $x = a$, but those values are not in these solutions. Here's what we can do to build the ground state wave function.

```
In [14]:   1  psi0=[0]
           2  for tpsi in psi.T[0]:
           3      psi0 = psi0 + [tpsi]
           4  psi0 = psi0 +[0]
           5  print(psi0)

[0, 0.1381966011250101, 0.26286555605956624, 0.3618033988749888, 0.4253
2540417601994, 0.4472135954999578, 0.4253254041760204, 0.36180339887498
95, 0.26286555605956713, 0.13819660112501073, 0]
```

The first step is to make a new list for this wave function. It starts off with the first value equal to zero (that's the first boundary condition). The for-loop then goes through the values in the transpose of the first eigenvector and adds each element to the psi0 list. Finally, one more zero is added for the other boundary condition. The eigenvector is printed just so you can see that it works.

Now we need to plot this solution. Here's the code using matplotlib.

In [17]:
```
1  plt.plot(x,psi0)
2  plt.xlabel("x")
3  plt.ylabel("psi")
4  plt.grid()
5  plt.show()
```

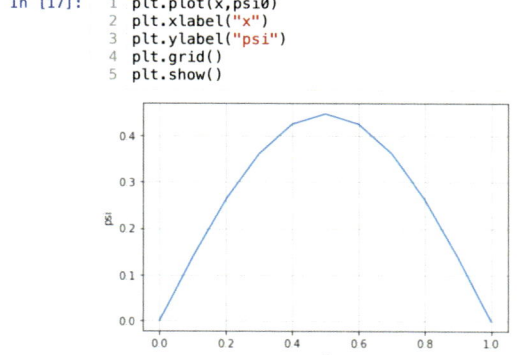

There are many options for creating plots with matplotlib.pyplot–but I have shown just the most common features. The plt.plot(x,psi0) just creates a plot with the values of x on the horizontal axis and psi0 on the vertical axis. The other options I used are xlabel and ylabel (to label the axis) and grid() to display grid lines.

The important point is that this does indeed look like a valid solution. However, it is not normalized. Let's normalize this function. Remember that the complex conjugate of the wave function gives the probability density. For our infinite square well, the particle must be somewhere between 0 and a. This means that the integral of the probability density over this region must be equal to 1.

$$\int_0^a \psi^*\psi dx = 1$$

In this example, we only have real values of the wave function, so we don't need to worry about the complex conjugate. We can just integrate over the length of the well. Since we have psi at different values of x, this integral just becomes a sum.

In [32]:
```
1
2  A = 0
3  for psit in psi0:
4      A = A + np.abs(psit**2*dx)
5  print("A = ",A)
6  for i in range(len(psi0)):
7      psi0[i] = psi0[i]/np.sqrt(A)
8  plt.plot(x,psi0)
9  plt.show()
10
11
```

A = 0.09999999999999995

In order to calculate a numerical integration, we start off with the variable $A = 0$. This will be the total "area" under the curve. Next, a for loop is used to go through all the values of psi0 and calculate the area between points. Notice that we are using the absolute value function to insure that we have a positive area. Finally, this area element is added to the total area. In this example, that area is 0.0999 and not 1.

We can force the wave function to be normalized by dividing each value by $1/\sqrt{A}$. Finally, this new wave function is plotted. Here's the normalized function.

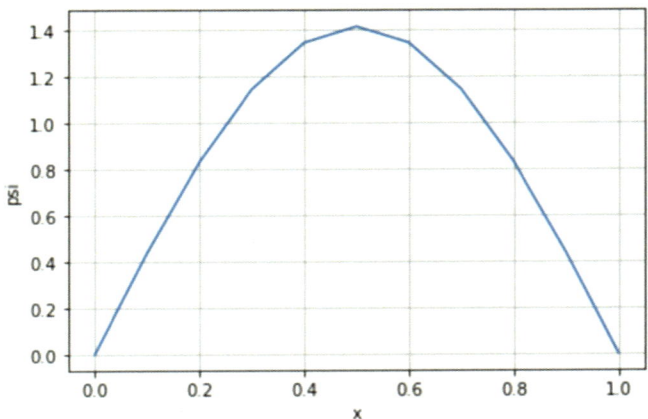

7.4   Infinite Square Well with a Step Potential

Now that we have a robust numerical method to solve the TISE, let's apply it to a situation that's not so trivial to solve analytically. Imagine that we have an infinite square well, but the "floor" isn't flat. Instead, there's a step on one side. It looks like this:

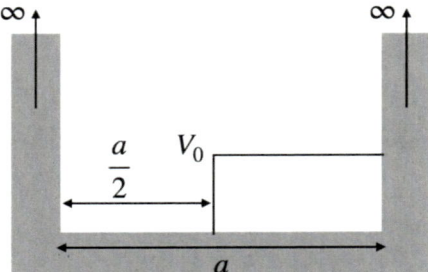

The only difference with this well and the flat one is that we have a non-zero potential inside the well. But we still need to solve the same time independent Schrodinger equation:

$$-\frac{\hbar^2}{2m}\frac{\partial^2\psi}{\partial x^2} + V(x)\psi = E\psi$$

If we break ψ into finite elements, then the second partial derivative can be replaced with the finite difference calculation. For the i-th element, we get the following equation.

$$-\frac{\hbar^2}{2m}\frac{(\psi_{i-1} - 2\psi_i + \psi_{i+1})}{\Delta x^2} + V_i\psi_i = E\psi_i$$

We can again make this into a tridiagonal matrix–but this time, there's an extra term in the diagonal from the potential energy term. Let's build this Hamiltonian matrix in order to calculate the eigenvectors. Here's the set up.

```
In [2]:    1  hbar = 1
           2  m = 1
           3  a = 1
           4  V0 = 10
           5  N = 100
           6  x = np.linspace(0,a,N+1)
           7  dx = x[1]-x[0]
           8  xp = np.linspace(dx,a-dx, N-1)
           9  V = np.zeros(N-1)
          10
          11  for i in range(len(V)):
          12      if dx*(i+1)>a/2:
          13          V[i]=V0
          14
          15  k = hbar**2/(2*m*dx**2)
```

We again have the same constants as with the normal square well. However, we need to define our potential function (V). If we use the x arrary to calculate our values of V, then there's going to be two extra items for the boundary conditions. This would be a problem since our Hamiltonian matrix doesn't have wave function values for the end points.

One solution is to make a new set of x values (called xp) that starts at x = dx and goes up to (but not including) x = a. This is what is done in line 8 above. Next, we can start with a potential function that has all zeros (using np.zeroes from before–as seen in line 9). Now I can go through theses zeroes of the potential and select which ones are changed. The for-loop in line 11 goes through the list. Line 12 checks if the x value is greater than a/2 and if so, it changes the potential value to V0.

Now we need to build the Hamiltonian matrix. Just for simplicity, I'm going to create it in three different steps as shown below.

```
In [6]:    1  H = np.diag(2*k*np.ones(N-1)+V)
           2  H = H - k*np.diag(np.ones(N-2),1)
           3  H = H -k*np.diag(np.ones(N-2),-1)
           4
```

The first step is to create the diagonal values. Notice the +V in this term–that's what makes it different from the previous square well. In line 2, I create the upper off diagonal.

Notice that I just add this off diagonal to the original H matrix. Finally, the lower off diagonal is created.

The eigenvalue problem is now just one line.

```
In [7]:   1  E,psi = np.linalg.eigh(H)
          2
```

That's the same as for the previous example. Notice that I'm not printing out the eigenvectors since I am using N = 100 (it would be too big). For the normalization, I once again need to integrate over the probability density. This is a little different than the last integration. In this case, I left off the end points for the wave function so that my psi values are actually numpy arrays instead of a python list. With that I can integrate with just one line.

```
In [8]:   1  A = np.sum(np.abs(psi.T[0])**2*dx)
          2  psi0 = psi.T[0]/np.sqrt(A)
          3
          4  plt.plot(xp,psi0**2)
          5  plt.xlabel("x")
          6  plt.ylabel("psi*psi")
          7  plt.grid()
          8  plt.show()
```

Line 2 divides the first eigenvector by the square root of the integration and makes it normalized. Plotting the probability function looks like this.

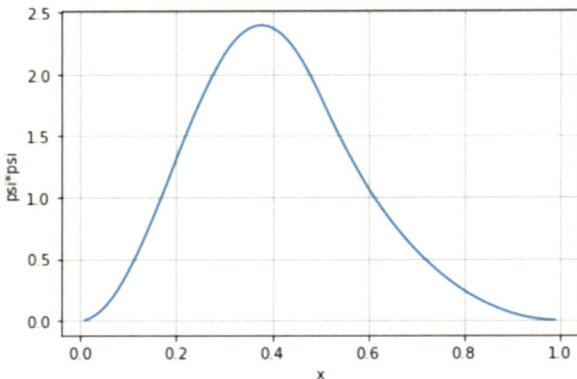

Notice that this looks similar to the plain 1D infinite square well–but slightly different. The probability density has a peak shifted towards x = 0. This actually makes sense. Since the right side of the well has a step, the particle is more likely to be found at x values less than a/2.

Just for fun, here are the probability densities for the second and third energy level.

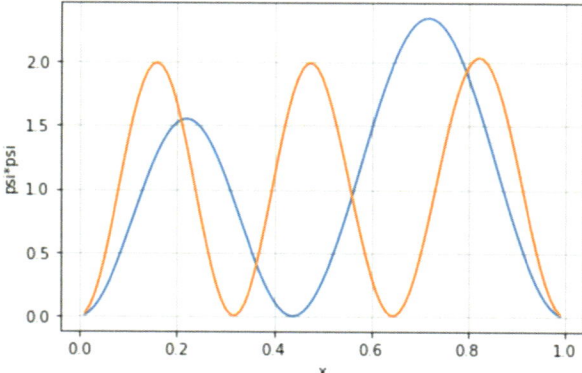

For the second energy level, the probability is higher for x > a/2 instead of less than a/2. We can see why this happens if we also print out the eigenvalues (the energy values). The energy value for the second eigenvector is 25.577 units. This is higher than the energy level of the step size (it was set at 10 units). You can think of this like a particle rolling up on a flat hill (when on the step). Since it's at a higher potential, the kinetic energy is lower. This means that the particle is effectively moving slower and more likely to be found there.

The third energy level is at 49.158 units. This is significantly higher than step energy so that the probability density mostly looks like the step is not even there. I's always nice to see if your numerical result make some sort of sense.

7.5   Half Finite Square Well

We've looked at two methods to solve the 1D TISE–the shooting method and the finite difference method. Which one is better? What are the advantages and disadvantages of each method? Let's consider another situation and solve it two different ways.

In this case, we are going to look at a half-infinite square well. This is similar to the infinite square well except that one of the walls isn't actually at an infinite potential.

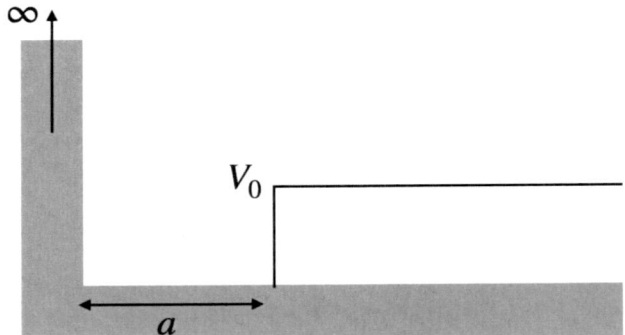

Since the right wall isn't at an infinite potential, it's possible to have solutions for the wave function in the region where x > a. Let's set this up with the shooting method using similar methods as in a previous code. Really, the only difference is the inclusion of the potential energy function (which will not always be zero).

$$-\frac{\hbar^2}{2m}\frac{\partial^2\psi}{\partial x^2} + V(x)\psi = E\psi$$

Remember that with the shooting method, we are going to start at x = 0 and use the Euler method to determine the values for ψ for the next value of x (using space step sizes of Δx). During each of these steps, we are going to assume that the second derivative of ψ (which we will call ψ'') is constant. Taking the above differential equation, we can solve for the second derivative as:

$$\psi'' = -\frac{2m}{\hbar^2}\psi(E - V)$$

If we assume ψ'' is constant, we can write it as:

$$\psi'' = \frac{\Delta\psi'}{\Delta x} = \frac{\psi'_2 - \psi'_1}{\Delta x}$$

where ψ'_1 is the value of the derivative for the first value of x and ψ'_2 is the derivative at the end of the intervale Δx. If the well has the infinite wall side at x = 0, then we know the value of ψ_1, but we don't actually know ψ'_1. We will just have to pick a value (we can use 1 because it's fun). With this, we can calculate ψ'_2

$$\psi'_2 = \psi'_1 + \psi''\Delta x$$

We will also make the assumption that ψ' is constant over this space interval (Δx) such that we can write it as:

$$\psi' = \frac{\Delta\psi}{\Delta x} = \frac{\psi_2 - \psi_1}{\Delta x}$$

From this we, can determine the value of ψ on the other side of the space interval (Δx).

$$\psi_2 = \psi_1 + \psi' \Delta x$$

The other problem is that we don't know the value of the energy (E). Again, we can just pick a value for that (starting at E = 0). This process is repeated for the next space interval until we get where we want (the other side of the well).

But there is something very different for this half-finite well. With the infinite square well, we knew where to "stop"–at the other side of the well. However, with this half infinite, we theoretically need to calculate all the way to x = infinity. Of course, that's not going to happen. We can compromise by going to x = 2a (twice the width of the well). Using the same methods as before, we get the following plot of the wave function.

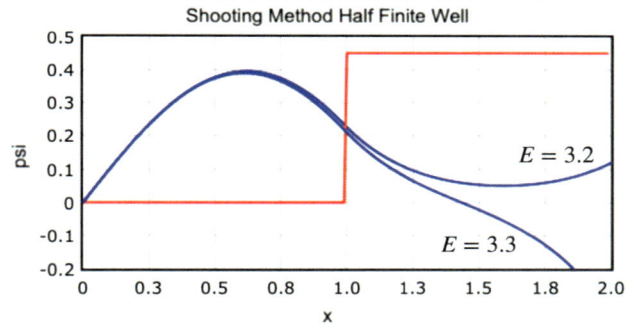

With this, we get an energy of 3.226 energy units. That's nice. But here's the issue– look at what happens when the energy is slightly off this value. Suppose I use an energy of 3.2 units and then 3.3 units. Here are those two wave functions.

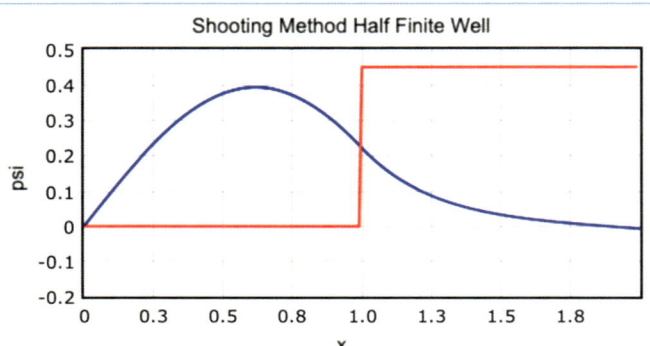

When the energy value is too low, the wave function explodes in the positive direction. When the energy is too high, the wave function explodes in the negative direction. So, the numerical solution flops from one extreme to the other. Of course, you can solve this problem by using a smaller energy step to get the value of energy that is well behaved. Another way to fix this problem is to calculate the wave function to a greater value of x.

Let's repeat this calculation by calculating the eigenvalues using the finite difference method.

Here is a plot of the first energy level.

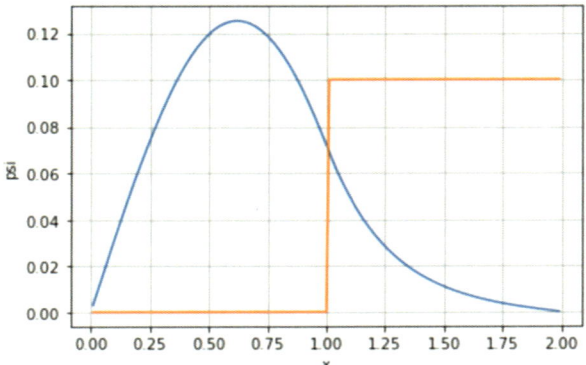

Using $N = 200$ finite values, this gives an energy of 3.1966 units which is fairly close to the value from the shooting method. The big difference is that the value of ψ can not explode at $x = 2a$ like with the shooting method. In this case, we have set the boundary condition of $\psi(2a) = 0$ as part of the Hamiltonian matrix.

So, which method is better? Clearly the eigenvalue method gives a better result. However, from a learning perspective, the shooting method is much easier to understand as it solves a differential equation in a method that's very similar to numerical method to determine the motion of a mass on a spring. For students that are learning numerical methods–the shooting method is much easier to pick up. On top of that, it doesn't require any specialized modules or functions (like the ones used in numpy) so that a student could even calculate this on paper using a calculator.

In the end, both methods are useful in their own way.

Time Evolution of the Wave Function

8

In the problems we have considered so far, we have assumed that the potential energy function only depends on space (x) and that our wave function had a quantized energy value. From this we had the following generic solution for the wave function.

$$\Psi(x, t) = \psi(x)e^{-iE_n t/\hbar}$$

Suppose I have an energy with n = 1. We can call this $\Psi_1(x, t)$. With this, we can calculate the probability density.

$$P(x, t) = \Psi_1^* \Psi_1$$

Recall that for the complex conjugate, we replace i (the imaginary number) with $-$ i. With this, we have the following.

$$P(x, t) = \psi_1^*(x)e^{iE_1 t/\hbar}\psi_1(x)e^{-iE_1/\hbar}$$

$$P(x, t) = \psi_1^*(x)\psi_1(x)e^{iE_1 t/\hbar - iE_1 t/\hbar} = \psi_1^*(x)\psi_1(x)$$

Because of the complex conjugate, the two terms with time cancel and we get a function that only depends on space (x). Since it doesn't change with time, we call this a stationary state.

It's also possible to have a combination of energies as a solution. Let's consider the infinite 1D square well (since we know the analog solution for this situation). Recall that the space part of the wave function would be:

$$\psi(x) = A \sin(kx)$$

where

$$k = \frac{n\pi}{a} = \frac{\sqrt{2mE}}{\hbar}$$

Such that the energies would be quantized as:

$$E_n = \frac{n^2\pi^2\hbar}{2ma^2}$$

If $\Psi_n(x, t)$ is a solution, then a linear combination of wavefunctions will also be a solution. Suppose we combine energies for n = 1 and n = 2. We can write the solution as:

$$\Psi(x, t) = c_1\Psi_1 + c_2\Psi_2$$

where c_1 and c_2 are constants. Let's write the wave functions with the time component included.

$$\Psi(x, t) = c_1\psi_1(x)e^{-iE_1t/\hbar} + c_2\psi_2(x)e^{-iE_2t/\hbar}$$

Let's again find the probability density.

$$P(x, t) = \Psi^*\Psi = \left(c_1\psi_1(x)e^{iE_1t/\hbar} + c_2\psi_2(x)e^{iE_2t/\hbar}\right)\left(c_1\psi_1(x)e^{-iE_1t/\hbar} + c_2\psi_2(x)e^{-iE_2t/\hbar}\right)$$

Expanding this gives:

$$P(x, t) = c_1^2\psi_1^2(x) + c_2^2\psi_2^2(x) + c_1c_2\psi_1(x)\psi_2(x)\left(e^{-i(E_2-E_1)t/\hbar} + e^{i(E_2-E_1)t/\hbar}\right)$$

Using the Euler identity (one of the many Euler identities) we can write the sum of exponentials as a trig function.

$$P(x, t) = c_1^2\psi_1^2 + c_2^2\psi_2^2 + c_1c_2\psi_1\psi_2 2\cos(\omega t)$$

where we can define ω as:

$$\omega = \frac{E_2 - E_1}{\hbar}$$

But what does this all mean? It means that if you have a system with a linear combination of wave functions, the probability density DOES depend on time. Also, the probability density oscillates with frequency of

$$f = \frac{\Delta E}{h}$$

where h is Plank's constant. Yes, this is the classic relationship between changes in energy and frequency of oscillation. Reminder–we made the assumption that the potential ONLY depends on space (x) and not time (t).

8.1   Modeling a Linear Combination of 2 States

We can model this 2 state oscillation in Web VPython. Suppose there is an infinite square well with a linear combination of wave functions for $n = 1$ and $n = 2$. Since we know the solutions for $\psi_n(x)$ we can write the total wave function as:

$$\Psi(x, t) = c_1 \psi_1(x) e^{-iE_1 t/\hbar} + c_2 \psi_2(x) e^{-iE_2 t/\hbar}$$

Just for this demonstration, I will pick the following values for c_1 and c_2:

$$c_1 = \sqrt{\frac{3}{10}}$$

$$c_2 = \sqrt{\frac{7}{10}}$$

I will use the same constants for the size of the well and mass of the particle as we used before. Here are the steps to make an animated graph of the probability density.

```
 3  gl = graph(xtitle="x [m]", ytitle="P", width=500, height=250,
 4  ymax = 1.4)
 5  f1 = gcurve(color=color.blue)
 6
 7  a = 1
 8  m = 1
 9  hb = 1
10  c1 = sqrt(3/10)
11  c2 = sqrt(7/10)
12  E1 = pi**2*hb**2/(2*m*a**2)
13  E2 = 4*pi**2*hb**2/(2*m*a**2)
14
```

There's really nothing new here that you haven't seen before. For the probability density, it changes with both time (t) and position (x). It's going to help if we create a python function to calculate the probability.

```
25  def P(xt,tt):
26      psi1 = c1*sin(pi*xt/a)
27      psi2 = c2*sin(2*pi*xt/a)
28      Ptemp = psi1**2+psi2**2+psi1*psi2*cos((E2-E1)*tt/hb)
29      return(Ptemp)
30
```

This uses the functions for the infinite square well from before. Now we just need to start with a time of t = 0 and go through all the values of x to calculate the probability density. After that, we move forward in time and repeat the whole process.

```
32  x = 0
33  dx = 0.01
34  t =0
35  dt = 0.01
36  while t < 5:
37      rate(100)
38      data = []
39      x = 0
40      while x<a:
41          data = data + [[x,P(x,t)]]
42          x = x + dx
43      fl.data = data
44      t = t + dt
```

Recall that for an animated graph, we add the data points (x an P) to a list and then plot all the points at one time before starting over. Also, I'm sure you remember that the rate (100) means that the code will not run any more than 100 loops per second. If you want the animation to run faster, increase that number.

When running, the graph will show the probability oscillating between these two forms.

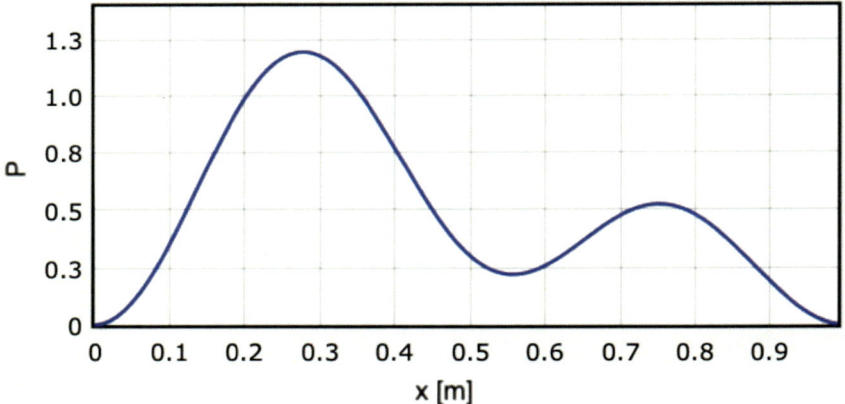

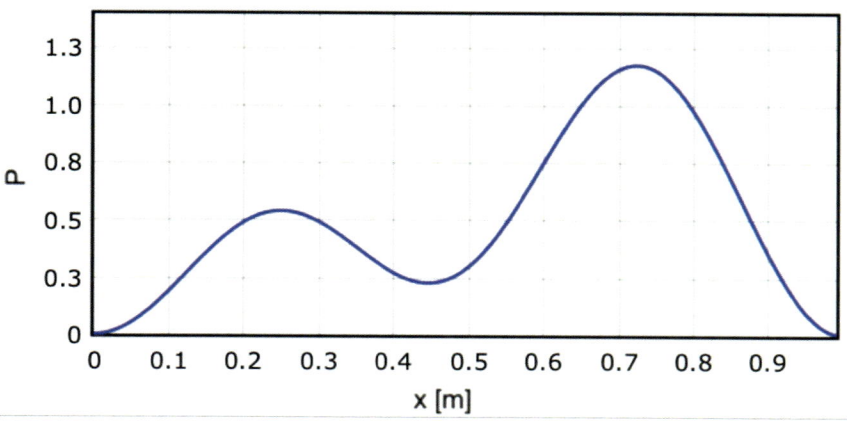

8.2   Evolution of an Initial Wave Function

Suppose we have our infinite square well and we know the initial shape of the wave function but we don't know the particular energy eigenvalues? How can we model the evolution of that system?

It's possible to represent a wave function as a linear combination of known solutions ($\psi_n(x)$) as before. It's possible that we could have an infinite number of solutions and a linear combination of these solutions.

$$\Psi(x, t) = \sum_{n=1}^{\infty} c_n \psi_n(x) e^{-iE_n t/\hbar}$$

However, then we will need an infinite number of coefficients c_n. We can calculate these using the state of the wave function at $t = 0$ and the known solutions.

$$c_n = \int_{-\infty}^{\infty} \psi_n^*(x) \Psi(x, 0) dx$$

Of course, for our infinite square well, the solutions are zero except for between $x = 0$ and $x = a$, so that makes the integration a little simpler. Also, for the infinite square well, we know the solutions for the normalized space part of the wave function.

$$\psi_n(x) = \sqrt{\frac{2}{a}} \sin\left(\frac{n\pi}{a} x\right)$$

Also (we will need this later), the energies are:

$$E_n = \frac{n^2 \pi^2 \hbar}{2ma^2}$$

Once we calculate the coefficients (c_n) and the energies for these values of n, we can determine the time portion of the wave function.

$$f(t) = e^{-iE_n t/\hbar}$$

But how do you model that in python since it's a complex number? We can again use the Euler identity:

$$e^{-i\theta} = \cos\theta - i\sin(\theta)$$

$$e^{-iE_n t/\hbar} = \cos(E_n t/\hbar) - i\sin(E_n t/\hbar)$$

With this (and assuming a real space function, $\psi_n(x)$) the full wave function can be split into a real (R) and imaginary component.

$$\Psi(x, t) = R(x, t) + iI(x, t)$$

With this, we can calculate the probability density function.

$$P(x, t) = \Psi^*\Psi = (R - iI)(R + iI) = R^2 + I^2$$

Here you can see that although the wave function can (and will be) a complex function, the probability density is a real function. This makes sense because the probability is a real value. Of course, the wave function still needs to be normalize such that:

$$\int_{-\infty}^{\infty} \Psi^*\Psi dx = 1$$

With all of this, we have the following recipe to model the evolution of a wave function with a known initial state (for the infinite square well). First, we need an initial function $\Psi(x, 0)$ either as numerical values or as a mathematical function. We need the known solutions for the situation ($\psi_n(x)$) and the energy values E_n. The second step is to take that initial shape and integrate to find the coefficients (c_n) we don't need an infinite number of coefficients, we can just calculate the first few (maybe 6 or 8 of them). With these coefficients, we can write the full wave function as sum of the first 6 or 8 solutions and add in the time component.

Here's the situation we wish to model. I'm going to use an infinite square well with width a = 6 units. The initial wave function at t = 0 will be:

$$\Psi(x, 0) = Ax(x - a)e^{-(x-a/2)^2}$$

We can start with a value A = 1 as a constant. The first step is to normalize this function by finding the value of A such that the integral of $\Psi(x, 0)^*\Psi(x, 0)$ is equal to 1. Of course we can do this numerically and it will be a nice warm up for later exercises.

Here's the starting stuff.

```
 9  dx = 0.1
10  x0 = 0
11  a = 6
12  hbar = 1
13  m = 1
14  t = 0
15  dt = 0.01
16
17  x = []
18  psi0 = []
19  xt = 0
20  A = 1
21  while xt<a:
22      x = x + [xt]
23      psi0 = psi0 + [A*xt*(xt-a)*exp(-(xt-a/2)**2)]
24      xt = xt + dx
25
```

After setting the constant values (lines 9–15), I need to make a list of x-values (line 17) as well as a list for the initial wave function values (calling this psi0). Here, I'm using a temporary x-value to move between the infinite walls and I'm calling it xt (for temporary x). The value of A is initially set to 1, but we are going to change that.

In lines 21–24, the list of values for x and psi0 are created. Notice that in line 23, the calculated value of $\Psi(x, 0)$ is added to the list.

Next, we can create a numerical integration to determine the value of the total probability.

```
27  PT = 0
28  for i in range(len(x)-1):
29      PT = PT + psi0[i]**2*dx
30  for i in range(len(x)):
31      psi0[i]=psi0[i]/sqrt(PT)
32
33
```

Here, the variable PT is the "total probability". Line 28 goes through all the intervals (which is 1 less than the number of x values) and calculates $\Psi^*\Psi dx$ then adds this to the total. Lines 30–31 goes back through the values of psi0 and divides by the square root of this value PT. If you went back through and integrate again, you would get a PT value of 1–so, it's normalized.

Just to make sure things are working, let's plot the probability density at a time = 0 s. Notice that since t = 0, the time part of the wave function is just 1 and we don't need to worry about real and imaginary parts of the wave function. Here's what it looks like.

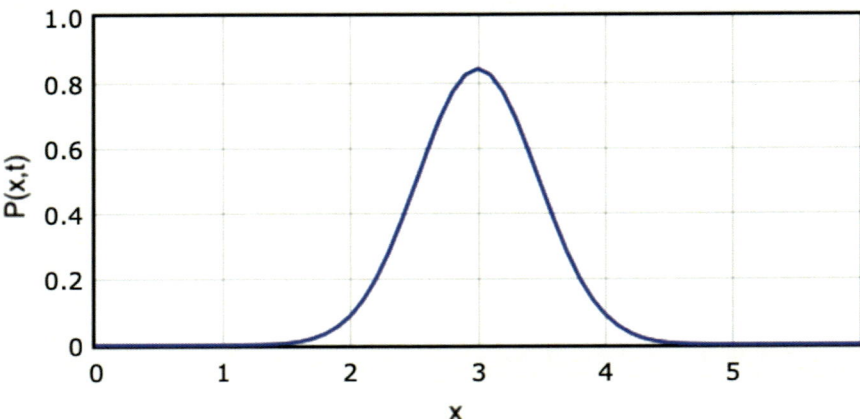

Now we are set to calculate our coefficients, c_n. Let's just calculate the first 8 coefficients (we clearly can calculate all infinity of them).

```
34  def E(nt):
35       return(nt**2*pi**2*hbar**2/(2*m*a**2))
36  def psin(xt, nt):
37       return(sqrt(2/a)*sin(nt*pi*xt/a))
38  cs = []
39  nn = 1
40  nmax = 8
41
42  while nn<nmax:
43     ct = 0
44     for i in range(len(x)-1):
45       ct = ct + psi0[i]*psin(nn,x[i])*dx
46     cs = cs + [ct]
47     nn = nn + 1
48  print(cs)
49
```

Just to make things simpler, I'm creating two python functions. The first one (E(nt)) calculates the energy value as a function of n. The second function (psin(nt,xt)) is the python version of $\psi_n(x)$ which I will need to calculate the c_n coefficients.

Next, I create a list for my values of c_n which I call cs. In lines 39 and 40 I create a counter variable (nn) and then the maximum number of coefficients to calculate (nmax).

In lines 42–47, we can calculate all these coefficients. The outer loop is just counting over the number of coefficients. Notice that since I'm going a numerical integration, I need to have a value for the sum of intervals. In this case, I'm calling that ct and it needs to be set to zero before the new calculation. Lines 44–45 are the numerical integration. The result is added to the list of coefficients.

Just to check that things seem to work, the values of the coefficients are printed. Here's what I get.

```
[-0.835146, 1.70326e-16, 0.513538, -6.61729e-17, -0.19215, -7.12607e-17,
0.0430651]
```

Notice that every other value is very close to zero, that's because of the symmetry of our initial function. Just to show you that this actually works, here is a reproduction of the initial wave function using just the first 6 coefficients (using 8 is too close to tell a difference).

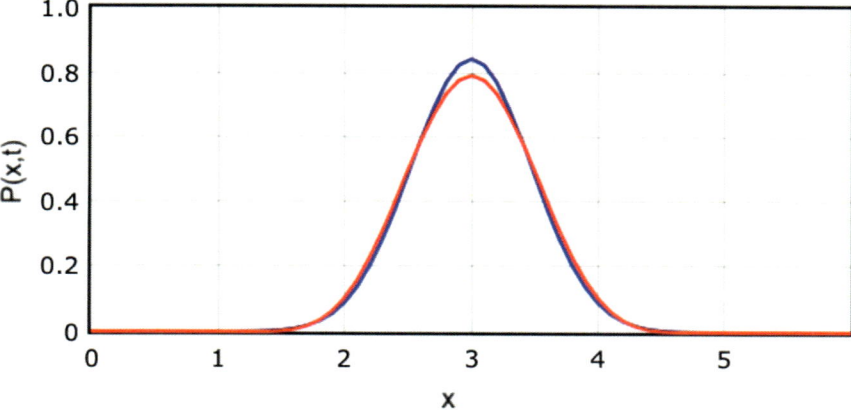

Now we can model the evolution of this wave function. Just to make things simpler, I'm going to create a couple more python functions—one for the real part of the wave function and one for the imaginary part.

```
66  def psiR(nt,xt,tt):
67      return(A*sin(nt*pi*xt/a)*cos(E(nt)*tt/hbar))
68  def psiI(nt,xt,tt):
69      return(A*sin(nt*pi*xt/a)*sin(E(nt)*tt/hbar))
70
```

Now I can calculate the value of $\Psi(x, t)$ using the real and imaginary functions and with that the probability density (which is a real function). Here you can see the output at three different times.

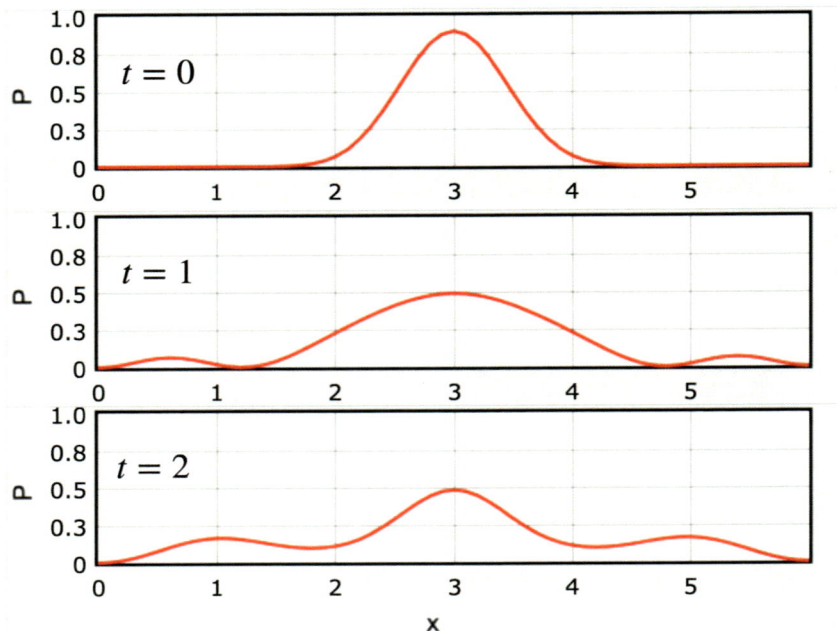

Although this method produces excellent results, it does have its limitations. In order to find the coefficients (c_n) and an expression for the wave function ($\Psi(x,t)$) we first need to know the eigenfunctions ($\psi_n(x)$). For an infinite square well, it's not too difficult to derive these functions. However, imagine that we have a square well with a step potential or maybe a half-finite well. In that case, we would need some other method to find not just the wave function for n = 1, but for at least the first few eigenfunctions.

8.3   Barrier Potential and Tunneling

Let's use the same idea for the evolution of a wave function to build a model to show what happens when a particle moves along in one dimension and encounters a barrier. If this barrier potential does not change with time, then just as before, we get the following function for the time part of the wave function.

$$f(t) = e^{-iE_n t/\hbar}$$

That leaves us with the following differential equation for the space part of the wave function.

$$-\frac{\hbar^2}{2m}\frac{\partial^2 \psi}{\partial x^2} + V\psi = E\psi$$

All we need is a function for the potential (V) and we can solve this problem using the finite difference definition of a second derivative and matrix operations (see the eigenvalue problem above). Of course, there are some issues that we need to deal with.

First, let's consider our barrier potential. Imagine this is a wall with an energy height of V_0 and a width of w. It would look something like this.

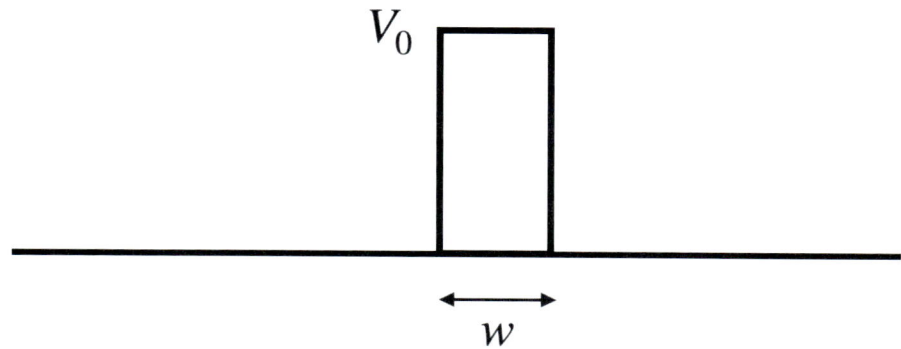

However, in our eigenvalue problem we need to have boundaries. For the infinite square well, it was simple to set the two infinite walls as locations where the wave function was required to be zero. If we want to make a particle interacting with the barrier, we don't have any obvious boundary conditions. But don't worry, we can just make some artificial boundaries. Basically, we can put this barrier in a very wide infinite square well such that we actually do know the boundary conditions.

Now we can just find the eigenfunctions for this square well potential. That's our next step. Let's go over the particular code for this calculation. Recall that since we need to find the eigenvalues of a matrix, we are going to have to use Jupyter notebooks instead of Web VPython.

The first step is to import the two modules that we will need.

```
In [5]:   1  import numpy as np
          2  import matplotlib.pyplot as plt
```

Now we need to declare our constants.

```
In [2]:    1  m = 1
           2  hbar = 1
           3  xmin = -6.5
           4  xmax = 6.5
           5  N = 1000
           6  x = np.linspace(xmin,xmax,N+1)
           7  dx = x[1]-x[0]
```

I'm again using weird (but easy) units for m and $\hbar$. For the infinite square well, I'm going to go from a value $- 6.5$ units to 6.5 units with N = 1000 intervals. That means that I will need N + 1 x values. So, in line 6 an array of x-values is created. Finally, I need the interval size dx (calculated in line 7).

The next step is to create the barrier. I'm going to make the barrier have an energy height equal to the kinetic energy of the particle.

```
In [6]:    1  p = 40
           2  V0 = p**2/(2*m)
           3  V = 0*x
           4  w = 0.5
           5  for i in range(len(V)):
           6      if x[i]>0 and x[i]<w:
           7          V[i] = V0
           8
           9  plt.plot(x,V)
          10  plt.xlabel("x")
          11  plt.ylabel("V")
          12  plt.show()
```

Here, p (in line 1) is the potential energy of the particle and that puts the max potential as:

$$V_0 = \frac{p^2}{2m}$$

Now, to create the actual values of the potential I'm going to use a small trick. In line 3 you can see V = 0 * x. Since x is a N + 1 length array, this makes V an array of the same size but with all the values equal to zero. I can just go through these values for V and switch some of them to V_0. Lines 4–6 go through each element in V. If the corresponding value of x is greater than 0 and less than w (the width of the barrier) then the element is set to V_0. Just for fun, I've included a plot of the potential.

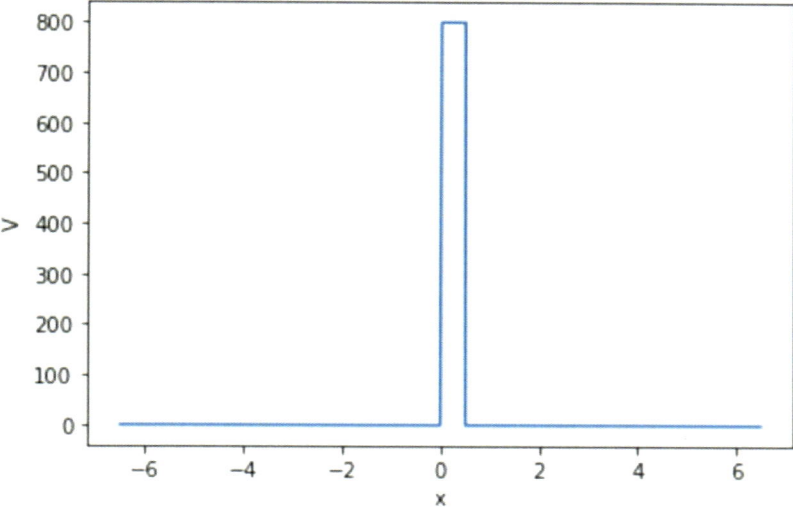

The next step is to create our initial wave function at $t = 0$ (just like in the previous example). Of course, in the infinite square well case the initial wave function was just a pulse–but it wasn't moving. We can create a moving pulse with the following function.

$$\Psi(x, 0) = e^{\frac{-(x-x_0)^2}{\sigma}} e^{ip(x+x_0)}$$

Because this is a complex number, it's going to make the pulse move to the right with a momentum p (you can see this by using the momentum operator, but that's not important right now). The pulse starts at a position x_0 and has a width of σ.

I can create this initial pulse and then plot the probability density at $t = 0$ (so that we won't have a complex number).

```
In [9]:    1  x0 = 2
           2  sig = 0.15
           3
           4  Psi0 = np.exp( -(x[1:-1]+x0)**2/sig**2)*np.exp(1j*p*(x[1:-1]+x0))
           5  AT = np.sum(np.abs(Psi0)**2*dx)
           6  Psi0 = Psi0/(np.sqrt(AT))
           7
           8  plt.plot(x[1:-1],np.abs(Psi0)**2)
           9  plt.xlabel("x")
          10  plt.ylabel("P")
          11  plt.show()
          12
```

There's an issue that we need to deal with. Just as before, we are going to create a Hamiltonian matrix with each dimension that's 2 less than the number of x values (since the boundary conditions are set). When calculating the value for the initial wave function (Psi0), we don't want to include the first or last x-value. The code x[1:−1] goes from the

element number 1 (which is the second element) up to (but not including) the last element
(−1).

Another issue is that the initial wave function is complex-it has an imaginary number
in there. In python, we can include imaginary numbers as "1j" that you can see in line 4
above.

This initial wave function needs to be normalized, so the integral over all space is
calculated in line 5 and set to the value AT. After that (in line 6) the values of Psi0
are divided by the square root of this value. Notice that when calculating the probability
density, we can take the absolute value of the function to remove the imaginary parts.

Here's a plot of that initial wave function.

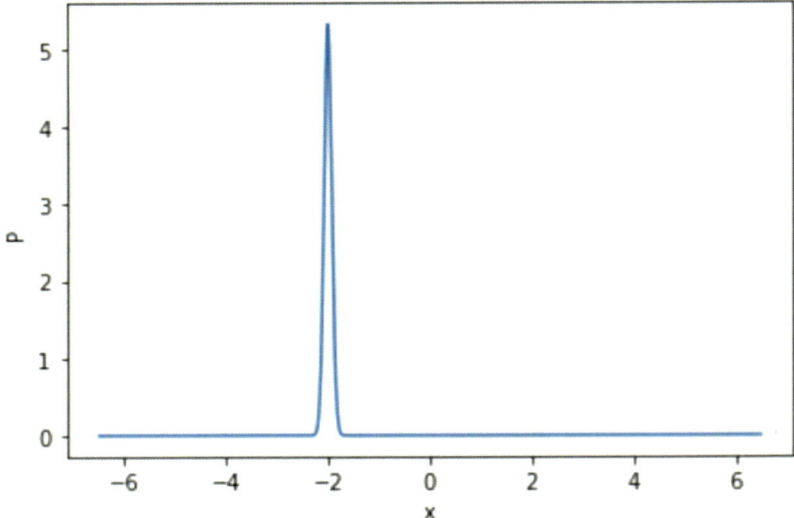

In order to model the time evolution of this initial wave function, we need to find the
eigenvectors ($\psi_n(x)$) for the system and then use them to calculate the coefficients c_n.
Just like in previous examples, this means that we will create a Hamiltonian matrix and
then find the eigenvalues and eigenvectors.

```
In [11]:   1  H = (hbar**2/(m*dx**2))*np.diag(np.ones(N-1))
           2  H = H + V[1:-1]*np.diag(np.ones(N-1))
           3  H = H + (-hbar**2/(2*m*dx**2))*np.diag(np.ones(N-2),1)
           4  H = H + (-hbar**2/(2*m*dx**2))*np.diag(np.ones(N-2),-1)
           5
```

Because the build is long, I put it in 4 different lines. The first line is the diagonal for
the second derivative. Line 2 adds the potential function to the diagonal. Lines 3 and 4
are the two off diagonal lines.

We can find the eigenvectors, normalize them and plot the first three solutions (just to check.

```
In [20]:   1  E,psi = np.linalg.eigh(H)
           2  psi = psi.T
           3
           4  A = np.sum(np.abs(psi[0])**2*dx)
           5  psi = psi/np.sqrt(A)
           6  plt.plot(x[1:-1],psi[0]**2)
           7  plt.plot(x[1:-1],psi[1]**2)
           8  plt.plot(x[1:-1],psi[2]**2)
           9  plt.xlabel("x")
          10  plt.ylabel("P")
          11  plt.show()
```

Here psi is an array of all the eigenvectors for this Hamiltonian matrix–but like before, we need to take the transpose of these solutions to get it in the form that's useful. Lines 4 and 5 normalize these functions. Finally, just as a check, I plot the probability density for the first 3 eigenvectors.

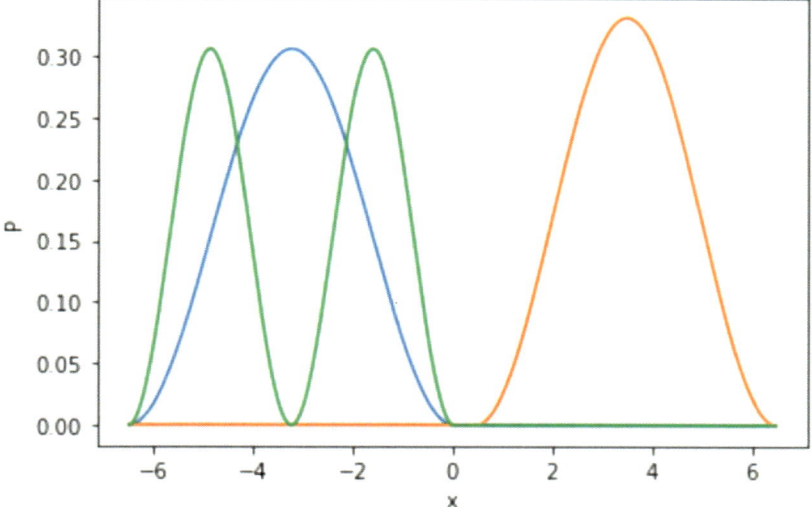

The first two energy levels are "trapped" on the left side of the barrier but the third energy level has enough energy to get past the barrier. Looks good.

Now we can calculate the coefficients (c_n). We actually don't have an infinite number of coefficients since we don't have an infinite number of eigenvectors (because we used the finite difference method). But I still need to find these values. There's a little trick. These coefficients can be complex, so I want to start off with them as complex numbers (in python). Since the array of Psi0 values is already complex, I can use this to create an array of complex numbers for the coefficients.

```
In [21]:    1  c = 0*Psi0
            2  for i in range(len(c)):
            3      c[i] = np.sum(np.conj(psi[i])*Psi0*dx)
```

After creating the array of values for c as complex numbers, lines 2–3 is just a python implementation of the integral calculation for the values of the coefficient. Notice that we are also taking the complex conjugate of the eigenvectors using np.conj.

With these coefficients, I can now calculate the full wavefunction for any given time as a sum of the space eigenvectors ($\psi_n(x)$) along with the time function.

$$\Psi(x, t) = \sum_{n=1}^{\infty} c_n \psi_n(x) e^{-iE_n t/\hbar}$$

The evolution of this wave function can now be modeled with the following steps. Start with an array of complex values for Ψ and use the values of c and the eignevectors (along with the eigenvalues for energy) to calculate $\Psi(x, t)$ for all values of x at that particular time (t). After that, we can change the value of time and repeat the process. Here's the code for this calculation.

```
In [23]:    1  t = 0
            2  dt = 0.02
            3  while t<0.1:
            4      Psi = 0*psi[0]
            5      for i in range(len(c)):
            6          Psi = Psi + c[i]*psi[i]*np.exp(-1j*E[i]*t/hbar)
            7      plt.figure(figsize=(6,1))
            8      plt.plot(x,0.001*V)
            9      plt.plot(x[1:-1],np.abs(Psi))
           10      plt.show()
           11      t = t + dt
```

Notice that the time step (dt $= 0.02$) is fairly large compared to the maximum time in the while loop (starting at line 3). This means that we are only going to calculate the wave function for 5 different times.

At the beginning of this loop of time, the values of Psi are set to a zero complex number. Lines 5–6 sum the coefficients (c) multiplied by the eigenvectors and the time part of the function. Notice that this uses the energy values E[i]. After that, the wave function is plotted along with a scaled down representation of the potential barrier.

Here is what the output looks like.

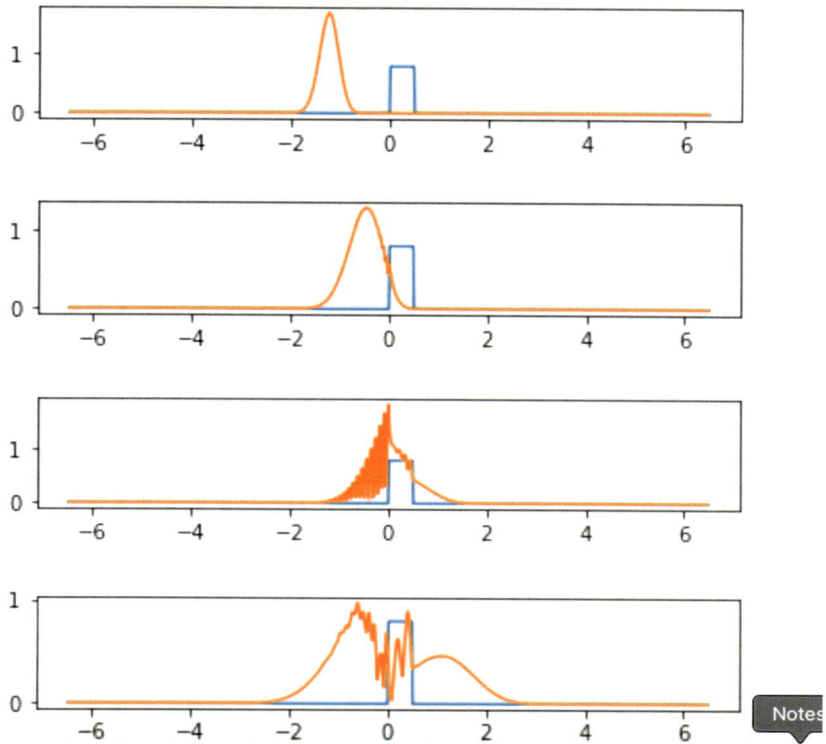

I'm displaying just the last 4 times to give you an idea of this evolution. Of course, it's also possible to create an animation–but this works fine.

Wave Functions in Two Dimensions

As you can see, even in 1 dimension quantum systems can get fairly complex. However, most of the real cases dealing with particles are in 3 dimensions. Of course the next best step is to look at a particle in 2D. Let's do that.

9.1 Particle in a 2D Infinite Square Well

The simplest 1D quantum system is an infinite square well. The same is true for 2D. We can think of a 2D infinite square well as an area in two dimensions with a particle confined in this region (a length of a in the x-direction and b in the y-direction).

In order to work in 2D, we need Schrodinger's equation in 2D. Recall the one dimensional version.

$$i\hbar \frac{\partial \Psi}{\partial t} = -\frac{\hbar^2}{2m} \frac{\partial^2 \Psi}{\partial t^2} + \Psi V$$

The partial derivative with respect to time is the same in 2D as it is in 1D, so we don't need to change that. For the space derivative, we need to replace this with the Laplacian. In Cartesian coordinates, we can write this operator as:

$$\nabla^2 = \left(\frac{\partial^2}{\partial x^2} + \frac{\partial^2}{\partial y^2} + \frac{\partial^2}{\partial z^2} \right)$$

Yes, that's 3D–but we can deal with the 2D version shortly. Using this operator, our Schrodinger equation becomes:

© The Author(s), under exclusive license to Springer Nature Switzerland AG 2025
R. Allain, *Modeling Waves with Numerical Calculations Using Python*, Synthesis
Lectures on Wave Phenomena in the Physical Sciences,
https://doi.org/10.1007/978-3-031-78291-6_9

$$ih\frac{\partial \Psi}{\partial t} = -\frac{\hbar^2}{2m}\nabla^2\Psi + V\Psi$$

Suppose that we write the wave function as a space part that only depends on x and y and a time part that only depends on t.

$$\Psi(x, y, t) = \psi(x, y)f(t)$$

Just as with the 1D version, if the potential does not depend on time we can separate the differential equation into time and space such that we get the following.

$$f(t) = e^{-iEt/\hbar}$$

$$-\frac{\hbar^2}{2m}\nabla^2\psi + V\psi = E\Psi$$

This means that we can find solutions to just the space part of the wave function (just like we did for the 1D square well).

For the space solution, we are going to make two assumptions. First that the potential is a linear combination of an x-potential and a y-potential. Second, that the wave function is a product of an x-function and y-function.

$$V(x, y) = V_x(x) + V_y(y)$$

$$\psi(x, y) = X(x)Y(y)$$

Applying the Laplacian to the wave function gives the following (note that there's no z-component in the wave function so that term just drops out).

$$\nabla^2(XY) = \frac{\partial^2 XY}{\partial x^2} + \frac{\partial^2 XY}{\partial y^2} = Y\frac{\partial^2 X}{\partial x^2} + X\frac{\partial^2 Y}{\partial y^2}$$

Now our TISE becomes:

$$-\frac{\hbar^2}{2m}\left(Y\frac{\partial^2 X}{\partial x^2} + X\frac{\partial^2 Y}{\partial y^2}\right) + (V_x + V_y)XY = EXY$$

Dividing both sides by XY:

$$-\frac{\hbar^2}{2m}\left(\frac{1}{X}\frac{\partial^2 X}{\partial x^2} + \frac{1}{Y}\frac{\partial^2 Y}{\partial y^2}\right) = E_x - V_x + E_y - V_y$$

Notice that I let $E = E_x + E_y$ just to make things work out in the next step. You can see that I have stuff that only depends on x and stuff that only depends on y. It's possible to get two equations from this—one dealing just with y and one dealing with x.

$$-\frac{\hbar^2}{2m}\frac{1}{X}\frac{\partial^2 X}{\partial x^2} = E_x - V_x$$

$$-\frac{\hbar^2}{2m}\frac{1}{Y}\frac{\partial^2 Y}{\partial y^2} = E_y - V_y$$

For our infinite 2D square well, both V_x and V_y are zero. That means our x-equation becomes.

$$\frac{\partial^2 X}{\partial x^2} = -\frac{2mE_x}{\hbar^2}X$$

Let's define the following.

$$k_x^2 = \frac{2mE_x}{\hbar^2}$$

Now this x-equation is in the same form for the 1D TISE for an infinite square well. We already know the solution for this differential equation.

$$X(x) = A\cos(k_x x) + B\sin(k_x)$$

Applying the boundary conditions that $X(0) = 0$ and $X(a) = 0$, we get the following.

$$X(x) = B\sin(k_x x)$$

$$k_x = \frac{n_x \pi}{a}$$

$$n_x = 1, 2, 3, \ldots$$

This gives us quantize values for the E_x energy (which is not the total energy).

$$E_x = \frac{n_x^2 \pi^2 \hbar}{2ma^2}$$

It should be straightforward to see that we can repeat this process for the y-equation to get:

$$Y(y) = B\sin\left(k_y y\right)$$

$$E_y = \frac{n_y^2 \pi^2 \hbar}{2mb^2}$$

There are some interesting features to this solution, however we need to remember why we found an analytical solution to begin with. The goal is to create numerical models for wave functions. By first creating an analytical solution, we can check our numerical methods to see if they are correctly working. However, in the end we have the following for the full wave function solution.

$$\Psi(x, y, t) = B \sin(k_x x) \sin(k_y y) e^{-iEt/\hbar}$$

Yes, I combine the two constants B into just one B. We can find this constant by normalizing the function. Again, this says that the total probability of finding the particle in the box must be 1.

$$\int\limits_{\infty}^{\infty} \int \Psi^* \Psi \, dx dy = 1$$

Notice that now we have a double integral since we are dealing with 2 dimensions. Using our wave function, we get:

$$\int\limits_{x=0}^{a} \int\limits_{y=0}^{b} B^2 \sin^2\left(\frac{\pi x}{a}\right) \sin^2\left(\frac{\pi y}{b}\right) e^{iEt/\hbar} e^{-iEt/\hbar} \, dx dy = 1$$

We can separate these into two integrals and integrate to obtain.

$$B^2 \frac{a}{2} \frac{b}{2} = 1$$

$$B = \frac{2}{\sqrt{ab}}$$

Now we have a full solution. Let's plot this solution for different values of n_x and n_y. Although I normally prefer to plot using Web VPython, for 2D graphs it's actually easier to use Jupyter Notebooks.

Of course the first step is to import the modules we will need (same as before).

```
In [1]:    1  import numpy as np
           2  import matplotlib.pyplot as plt
```

Next, we need to declare constants. In this case, I choose a = b = 1.

```
In [2]:    1  m = 1
           2  hbar = 1
           3  a = 1
           4  b = 1
           5  B = 2/np.sqrt(a*b)
```

In order to make a 2D plot in python we need to make a meshgrid. You can think of this as all of the possible combinations of our x and y values that we are going to plot. It's like the 2D version of our array of x values that we made in 1D.

Let's start with something simple–just a 10×10 grid (so we can print it see what's happening).

```
In [5]:   1  N = 10
          2  x = np.linspace(0,a,N+1)
          3  y = np.linspace(0,a,N+1)
          4  X,Y = np.meshgrid(x,y)
          5  print(X)
```

Notice that we first have to create two arrays, one for x and one for y–just like we did before. With these arrays, we combine them into a meshgrids (X and Y). If we print the X mesgrid, you can see what's happening. Note: don't print this for large meshgrids because they get huge.

```
[[0.   0.1 0.2 0.3 0.4 0.5 0.6 0.7 0.8 0.9 1. ]
 [0.   0.1 0.2 0.3 0.4 0.5 0.6 0.7 0.8 0.9 1. ]
 [0.   0.1 0.2 0.3 0.4 0.5 0.6 0.7 0.8 0.9 1. ]
 [0.   0.1 0.2 0.3 0.4 0.5 0.6 0.7 0.8 0.9 1. ]
 [0.   0.1 0.2 0.3 0.4 0.5 0.6 0.7 0.8 0.9 1. ]
 [0.   0.1 0.2 0.3 0.4 0.5 0.6 0.7 0.8 0.9 1. ]
 [0.   0.1 0.2 0.3 0.4 0.5 0.6 0.7 0.8 0.9 1. ]
 [0.   0.1 0.2 0.3 0.4 0.5 0.6 0.7 0.8 0.9 1. ]
 [0.   0.1 0.2 0.3 0.4 0.5 0.6 0.7 0.8 0.9 1. ]
 [0.   0.1 0.2 0.3 0.4 0.5 0.6 0.7 0.8 0.9 1. ]
 [0.   0.1 0.2 0.3 0.4 0.5 0.6 0.7 0.8 0.9 1. ]]
```

These are all our x values at the different points in 2D (as a 2D array). If you print the Y part, it will be rotated because we are dealing with the y values.

Next, let's create a python function that calculates the space part of our wave function.

```
In [8]:   1  def psi(nxt, nyt):
          2      return(B*np.sin(nxt*np.pi*X/a)*np.sin(nyt*np.pi*Y/b))
```

Remember, our wave function depends on 2 quantum numbers, n_x and n_y. When I pass these values into the wave function, I'm using temporary values nxt, and nyt. However, the function uses the global meshgrid values X and Y.

Now, to plot this function in 2D, we have some options. The most common is a contour plot.

```
In [9]:   1  plt.contour(X,Y,psi(3,2),levels=20)
          2  plt.xlabel("x")
          3  plt.ylabel("y")
          4  plt.show()
```

The contour plot takes the X and Y values of the meshgrid for the x–y values in the plot. The contours are created from the function psi–in this case I'm using the quantum numbers 3 and 2. The levels $= 20$ tells the plot how many lines to split for the contour. Here's what it looks like.

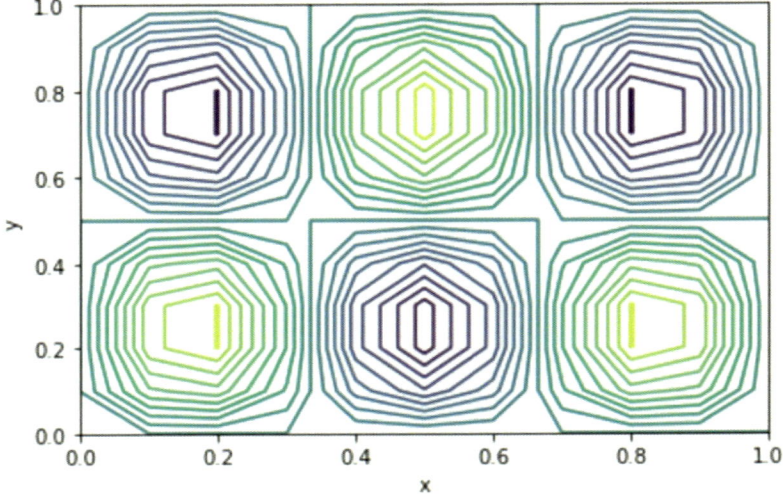

If we increase our value from N $= 10$ to N $= 100$, we get this output.

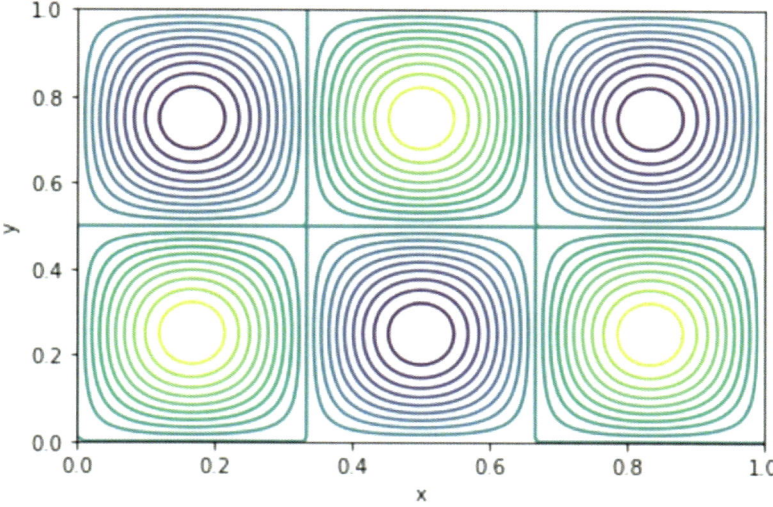

There are other ways to plot 2D data. Another popular method is the contour–this is the same except that the plot is filled in between the contour lines.

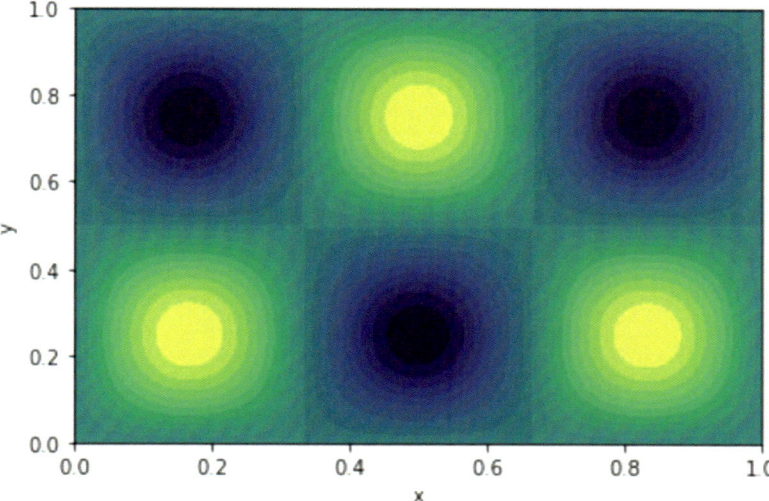

9.2   2D Wave Function with Non-trivial Potentials

The solution to the infinite square well doesn't require a numerical calculation. However, imagine that we have a 2D well with a bump in the bottom such that V is not zero everywhere inside the well. How do you solve a situation like this?

Let's recall the space part of the Schrodinger's equation.

$$-\frac{\hbar^2}{2m}\nabla^2\psi(x, y) + V\psi(x, y) = E\psi(x, y)$$

In 1D, we solved this equation by writing the partial derivative as a finite difference, but how do we do that with the Laplacian? Let's start off by again imagining that we break the wave function into finite elements in 2D. We can label each value of ψ with two indices–one for the x location and one for the y.

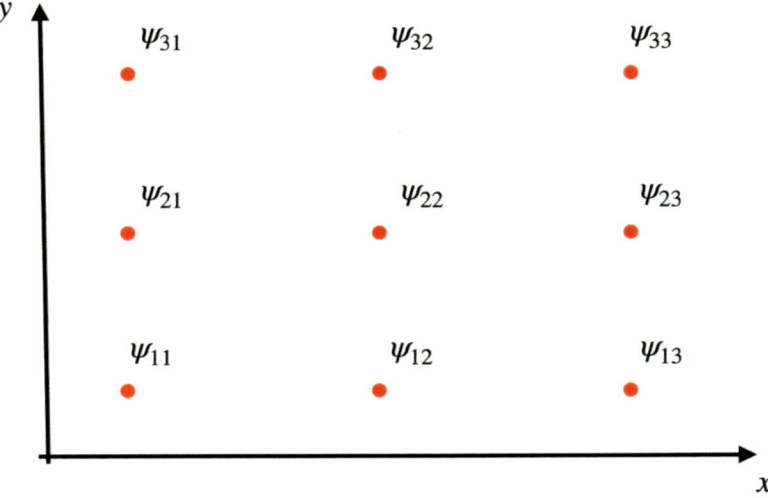

We can represent these finite elements as a matrix.

$$\psi = \begin{pmatrix} \psi_{11} & \psi_{12} & \psi_{13} & \cdots & \psi_{1N} \\ \psi_{21} & \psi_{22} & \psi_{23} & \cdots & \psi_{2N} \\ \vdots & \vdots & \vdots & \ddots & \vdots \\ \psi_{N1} & \psi_{N2} & \psi_{N3} & \cdots & \psi_{NN} \end{pmatrix}$$

However, if we want to use the same method of the eigenvalue problem for the 1 dimensional case then we need this to be a vector and not an N x N matrix. Of course we still need all the values–so we are going to make this a $1 \times N^2$ matrix.

$$\psi = \begin{pmatrix} \psi_{11} \\ \psi_{12} \\ \vdots \\ \psi_{1N} \\ \psi_{21} \\ \vdots \psi_{NN} \end{pmatrix}$$

Notice that this wave function vector starts off with the first row of the original matrix followed by the second row and so forth. We really just cut up this matrix and put it back together as one long series of values.

Although we have created a vector for our element values, we also caused a problem. Recall that in 1D we define the second derivative as the following finite difference.

$$\frac{\partial^2 \psi_i}{\partial x^2} = \frac{\psi_{i+1} - \psi_i - \psi_i + \psi_{i_1}}{\Delta x^2} = \frac{\psi_{i+1} - 2\psi_i + \psi_{i-1}}{\Delta x^2}$$

But this won't work now since the next element in the list might not be the one we want. Along with that, we actually need the Laplacian–at least the Laplacian in two dimensions. It is possible to create a matrix operator that actually does the derivative in a way that works. It looks like this.

$$D = \begin{pmatrix} -2 & 1 & 0 & 0 & \dots & 0 \\ 1 & -2 & 1 & 0 & \dots & 0 \\ 0 & 1 & -2 & 1 & \dots & 0 \\ 0 & 0 & 1 & -2 & \dots & 0 \\ \vdots & \vdots & \vdots & \vdots & \ddots & \vdots \\ 0 & 0 & 0 & 0 & \dots & -2 \end{pmatrix}$$

This looks like our second derivative from the 1D case. But how do we make this work with a partial derivative in the x-direction (using the first index of the ψ element) as well as a derivative in the y-direction (using the second index of the element)?

The answer is to use a Kronecker sum and the Kronecker product. These are matrix operations represented as $\oplus$ for the sum and $\otimes$ for the product. With that, we can combine the derivative matrix operator for x and y to represent the 2D Laplacian as:

$$\frac{\partial^2 \psi}{\partial x^2} + \frac{\partial^2 \psi}{\partial y^2} = D_{xx} \oplus D_{yy} = D_{xx} \otimes I + I \otimes D_{yy}$$

where D_{xx} and D_{yy} are the finite derivative matrices and I is the identity matrix. If you wanted to represent the Kronecker product for I and one of the D operators, it would look like this.

$$
I \otimes D = \begin{pmatrix} D & 0 & 0 & 0 & \dots & 0 \\ 0 & D & 0 & 0 & \dots & 0 \\ 0 & 0 & D & 0 & \dots & 0 \\ 0 & 0 & 0 & D & \dots & 0 \\ \vdots & \vdots & \vdots & \vdots & \ddots & \vdots \\ 0 & 0 & 0 & 0 & \dots & D \end{pmatrix}
$$

Here we have a matrix (it's huge) in which the diagonal elements are themselves matrices. What about the product of D and I? It's actually not the same thing (this operator does not commute) so that we get:

$$
D_{xx} \otimes I = \frac{1}{\Delta x^2} \begin{pmatrix} -2I & I & 0 & 0 & \dots & 0 \\ I & -2I & I & 0 & \dots & 0 \\ 0 & I & -2I & I & \dots & 0 \\ 0 & 0 & I & -2I & \dots & 0 \\ \vdots & \vdots & \vdots & \vdots & \ddots & \vdots \\ 0 & 0 & 0 & 0 & \dots & -2I \end{pmatrix}
$$

Again, we have I has the identity matrix for the diagonals and off diagonals. All of these terms have the $1/\Delta x^2$ term in them so that we can factor that out. But now we have a way to represent the 2D Laplacian using these very large matrices. We need to also represent the potential as a matrix but it should only have diagonal values–so that's fairly easy to create. We still need to take our 2D potential values and cut it up to spread along the diagonal. Here's what that looks like.

$$
VI = \begin{pmatrix} V_{11} & 0 & 0 & 0 & \dots & 0 \\ 0 & V_{12} & 0 & 0 & \dots & 0 \\ 0 & 0 & V_{13} & 0 & \dots & 0 \\ 0 & 0 & 0 & V_{14} & \dots & 0 \\ \vdots & \vdots & \vdots & \vdots & \ddots & \vdots \\ 0 & 0 & 0 & 0 & \dots & V_{NN} \end{pmatrix}
$$

Putting all of this together, our Schrodinger equation now takes the form:

$$
\left[-\frac{\hbar^2}{2m}(D \oplus D) + VI \right]\psi = E
$$

Just to recap (since this gets sort of confusing). In order to work in 2D, we take our two dimensional elements of ψ and spread them into a very long vector. The Laplacian is represented by a new (very large) matrix that's really a matrix with a bunch of matrices in them. The point is not that this is simple to do but rather that we have a method to actually turn this 2D problem into an eigenvalue problem.

9.3   Python Implementation

We are going to need some new tools and tricks to get this to work in python. Again, since we are dealing with matrices we are going to need to use Jupyter notebooks instead of Web VPython.

```
In [1]:    1  import numpy as np
           2  from scipy.sparse.linalg import eigsh
           3  from scipy.sparse.linalg import eigs
           4  import matplotlib.pyplot as plt
           5  from scipy import sparse
```

Other than numpy and matplotlib.pyplot, there are some new modules here. The scipy.sparse module has functions that let us deal with sparse matrices–these are matrices that are mostly empty (zeros for many of the elements). Since we are dealing with giant matrices, this will help things run faster.

Now we can start building stuff. Here are our constants along with our x and y values for the meshgrid.

```
In [2]:    1  m = 1
           2  hbar = 1
           3  N = 150
           4  a = 1
           5  b = 2
           6  x = np.linspace(0,a,N)
           7  y = np.linspace(0,b,N)
           8  X,Y = np.meshgrid(x,y)
           9  dx = x[1]-x[0]
          10  dy = y[1]-y[0]
```

To start off, we can create a potential function (V). I'm going to use $V(x,y) = 0$ everywhere since that will match our analytical solution for the 2D infinite square well. After we check that this works, we can move on to something more interesting.

```
In [6]:    1  def potential(xt,yt):
           2      return(0*xt+0*yt)
           3
           4  V = potential(X,Y)
           5
```

Here I created a function to calculate the values of V(x,y)–for now, it's just zeros. Then the actual potential (in line 4) uses the X and Y meshgrid variables so that we get a meshgrid of potentials.

Next, we need to build the matrix to represent a a finite difference derivative. We can then build our Hamiltonian as the sum of the kinetic energy operator (T) and the potential energy (which I will call U).

```
In [7]:    1  #make D
           2  diag = np.ones([N])
           3  diags = np.array([diag,-2*diag,diag])
           4  D = sparse.spdiags(diags,np.array([-1,0,1]),N,N)
           5
           6  #make T
           7  T = -hbar**2/(2*m)*sparse.kronsum(D,D)
           8
           9  #fix V
          10  U = sparse.diags(V.reshape(N**2),(0))
          11
          12  #the Hamiltonian
          13  H = T+U
```

The D operator starts off as a matrix of matricies. After that, we have the kinetic energy term and the potential term. Now we can find the energy eigenvalues and the eigenvectors.

```
In [8]:    1  E, eigenv = eigsh(H,k=10,which='SM')
```

The eigsh function returns two things—the energy values (E) and the eigen vectors. Since this is a HUGE matrix, you would get a HUGE number of energies and vectors. In this case, we are telling it to just get 10 values. Even with just 10 values, it still takes some time to run this part. I'm not sure what the which = 'SM' thing does, but you need it.

From this, we want the values of ψ at our finite locations (in 2D) but our solution is in squished vector form. We can un-squish the solution with the following function.

```
In [9]:    1  def psi(n):
           2      return(eigenv.T[n].reshape(N,N))
```

This function takes an energy value of n and returns a matrix of the unsquished solutions. We still need to normalize the values of ψ. Here's an example for the 6th energy level.

```
In [10]:    1  A = np.sum(np.abs(psi(6))**2*dx*dy)
            2  psi6=psi(6)/np.sqrt(A)
```

Since the psi function is a matrix, the np.sum actually acts as a two dimensional integral—which is exactly what we need. Just to check, let's plot the probability density for the n = 2 energy level.

```
In [13]:    1  plt.contour(X,Y,psi1**2, levels=20)
            2  plt.show()
```

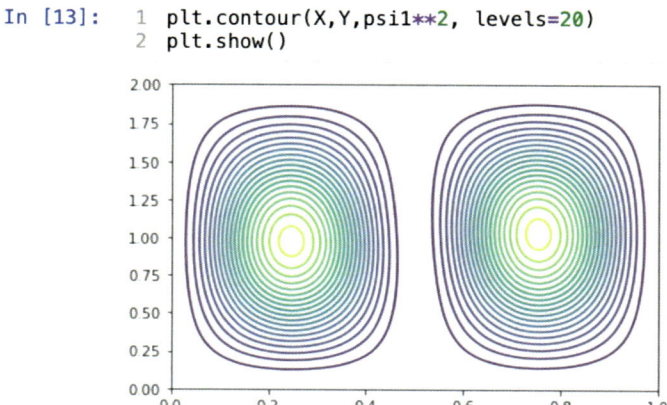

Now that this seems to work, let's move on to a non-trivial potential. Suppose we use a potential function that gives us a "bump" in the infinite square well.

$$V(x, y) = e^{-(x-x_0)^2/\sigma^2} e^{-(y-y_0)^2/\sigma^2}$$

Going back to our python code, we just have to change the definition of the potential energy function.

```
In [27]:    1  def potential(x,y):
            2      return(np.exp(-(x-0.3)**2/(4*0.1**2))*np.exp(-(y-0.3)**2/(8*0.1**2)))
            3  V = potential(X,Y)
            4  plt.contour(X,Y,V)
            5
```

Just to check, let's plot this potential. Here's what it looks like.

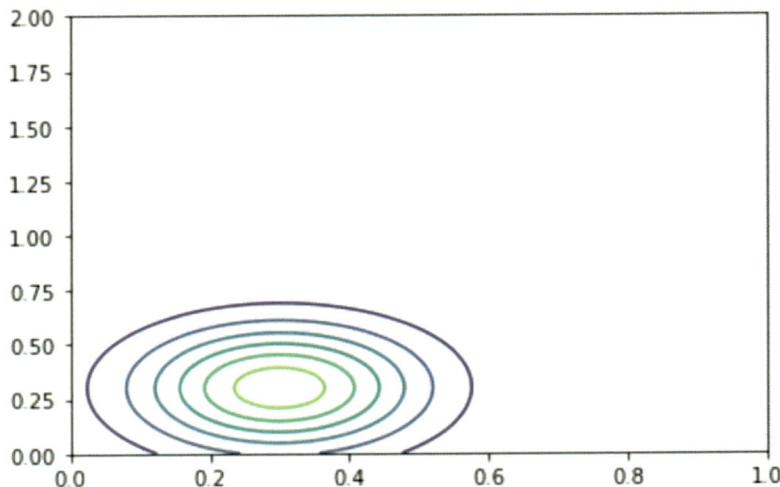

Now that we have a new potential, we can recalculate the eigenvectors. Here is a contour plot of the probability density for the first energy level.

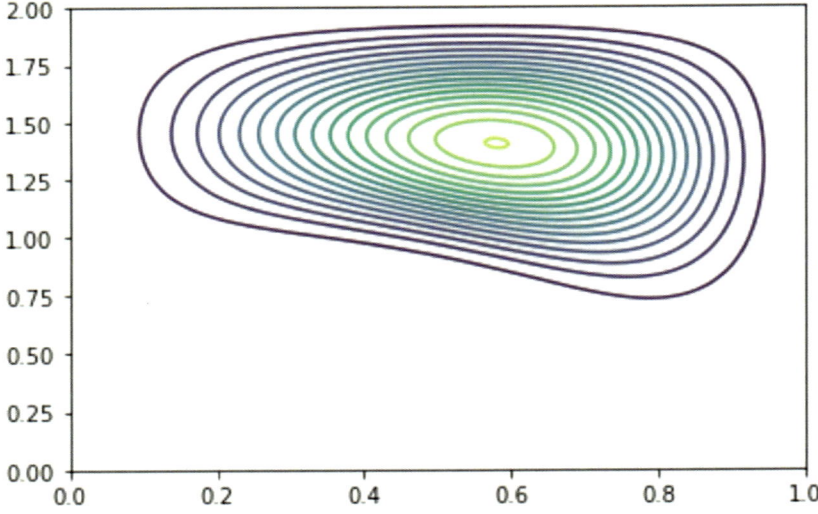

Notice that we are more likely to find the particle somewhere other than the lower left corner of the infinite square well since that's where the potential bump is to push the particle out of that region.

Let's try something with a higher energy level. How about the 6th level (which would be n = 5 in the code).

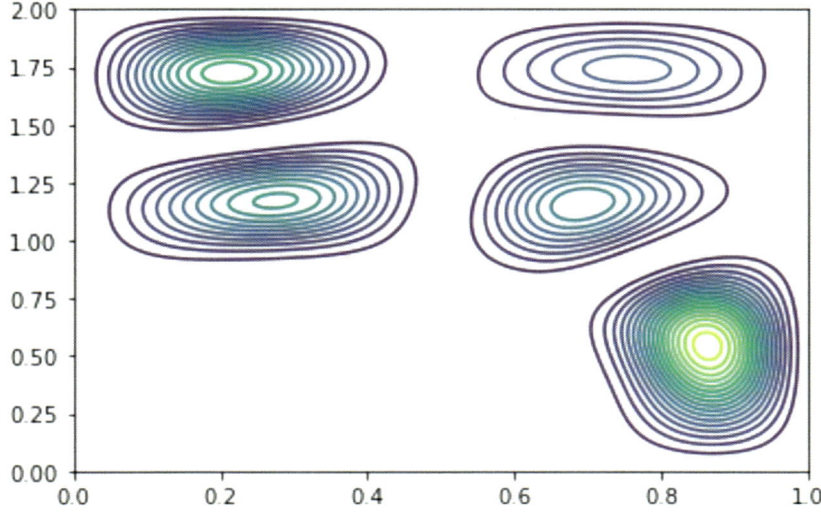

9.4   Summary and What's Next?

Take a moment to look back at what we have accomplished. We started off with a simple mass connected to a spring. Although this is a solvable problem, we were also able to model this with a numerical calculation. The key was to break the motion into short intervals of time–during those intervals, we assumed the acceleration was constant. This turned it from one complicated problem into many trivial problems and made it perfect for a computer program to calculate.

If you can model one mass on a spring, you can model many masses with many springs. This allowed us to look at the motion of a wave on a string and explore properties of waves (and the wave equation).

We then expanded this idea of waves to look at solutions to Schrodinger's equation. Although it's not actually a wave equation, it is a differential equation that changes with both space and time (just like a wave on a string). In the case of a potential function that does not depend on time, we can separate the wave function into a part that only depends on space (which we call ψ) and a part that depends on time. The space part of the wave function can be determined using the boundary conditions.

It's also possible to model the evolution of a wave function in time. The most important idea is to represent the second partial derivatives as a finite difference. With that, the Schrodinger equation can be represented as a matrix operation. This turns the equation into an eigenvalue problem–which we can solve even though it deals with very large matrices (with the help of python). Once we have eigenvector solutions to our potential well, we can evolve any linear combination of solutions to model many different wave functions.

Finally, we expanded this eigenvector idea to include situations in two dimensions. The basic process is the same as a 1D model, but the matrices get MUCH larger. It's possible to also look at the evolution of a 2D wave function using similar ideas that we used in 1D but you can see that things can get computational complex.

What about a 3D problem with Schrodinger's equation. Is that possible? Yes, we can use similar methods to determine probability densities for electrons near a nucleus for different atoms. I always like to point out how important numerical calculations have become. In an introductory quantum mechanics class, it's possible that you will analytically solve for the wave function due to the potential in a hydrogen atom (one proton and one electron). However, after that the problem becomes essentially impossible. You either have to make an approximation to solve for things like a helium atom or you have to do it numerically. We can't solve most of these equations without numerical calculations.